AF821138

de Gruyter Expositions in Mathematics 51

Editors

V. P. Maslov, Academy of Sciences, Moscow
W. D. Neumann, Columbia University, New York
R. O. Wells, Jr., International University, Bremen

State Observers for Linear Systems with Uncertainty

by

Sergey K. Korovin and Vasily V. Fomichev

Walter de Gruyter · Berlin · New York

Authors

Sergey K. Korovin
Faculty of Computational Mathematics
and Cybernetics
Moscow State University
Vorob'evy Gory
119899 Moscow, Russia
E-mail: korovin@isa.ru

Vasily V. Fomichev
Faculty of Computational Mathematics
and Cybernetics
Moscow State University
Leninskie Gory
119992 Moscow, Russia
E-mail: formichev@cs.msu.su

Mathematics Subject Classification 2000: 93-02, 93B07, 93B35, 93B50

Key words: System theory, control theory, state control, dynamical systems, systems of linear ordinary differential equations.

♾ Printed on acid-free paper which falls within the guidelines of the ANSI to ensure permanence and durability.

ISSN 0938-6572

ISBN 978-3-11-021812-1

Bibliographic information published by the Deutsche Nationalbibliothek

The Deutsche Nationalbibliothek lists this publication in the Deutsche Nationalbibliografie; detailed bibliographic data are available in the Internet at http://dnb.d-nb.de.

© Copyright 2009 by Walter de Gruyter GmbH & Co. KG, 10785 Berlin, Germany.
All rights reserved, including those of translation into foreign languages. No part of this book may be reproduced or transmitted in any form or by any means, electronic or mechanical, including photocopy, recording, or any information storage or retrieval system, without permission in writing from the publisher.

Printing and binding: Hubert & Co. GmbH & Co. KG, Göttingen.
Cover design: Thomas Bonnie, Hamburg.

Preface

In this book the authors show the latest (approximately twenty-year) achievements in the field of synthesis of state observers of dynamical systems described by ordinary differential equations or by recurrence relations with finite memory. The results of the preceding time period are reflected, sufficiently completely, in the monograph by O'Reilly.

The main achievements concern the development of the observability theory for multidimensional (multiply connected) systems, functional observers, and observers under the conditions of uncertainty. In addition, an essential progress was achieved in the synthesis of the simplest observers, i. e., minimal-order observers. This problem was investigated for standard as well as for functional observers. The main idea, which combines all problems, is the idea of obtaining the necessary information about a system with the use of minimal means.

One more problem touched upon in the book concerns statical and nonstatical methods of estimation under uncertainty conditions, algorithms of estimation which give an asymptotically exact reconstruction of a function or an estimation with an error which can be arbitrarily regulated.

The book is intended for specialists in the theory of automatic control as well as for lecturers, students, and post-graduates of the corresponding specialities.

When we wrote the book and prepared the manuscript, we received essential support from the senior lecturer of the Moscow State University A. V. Ilyin and the post-graduate of the State University I. S. Medvedev. We express our gratitude to them for their help and for the possibility of using the results of the joint investigations in this book.

The authors are also grateful to Academician Stanislav Vasilyevich Emelyanov for his help in the process of investigations and preparation of the manuscript.

S. K. Korovin, V. V. Fomichev Moscow 2009

Contents

Chapter 1

Notion of state observers

The problem of synthesis of state observers for dynamical systems, including automatic control systems, is a classical one and has rich history.

Everywhere in the sequel, for definiteness, by a dynamical system we mean a control system. In the finite-dimensional case for continuous time an automatic control system is described by a system of ordinary differential equations whose right-hand side depends on the *input of the system* $u(t)$ choosing which we can influence the properties of a given system. In the general form such a system is defined by a vector differential equation

$$\dot{x} = f(x, u, t), \quad t \geq 0, \tag{1.1}$$

where $x \in \mathbb{R}^n$ is a phase vector of the system. The necessity of a state observer is conditioned by the fact that when solving control problems we often have the information not about the phase vector x but only about a certain function of x,

$$y = h(x), \tag{1.2}$$

which is called the *output of the system* which, in general, makes it difficult to solve a control problem with the necessary correctness.

By the problem of constructing a state observer we understand the synthesis of a dynamic object which forms the estimate of the vector of states of the dynamical system with the use of the information that we have about the system, its measurable output and input.

A huge number of works are devoted to the solution of this problem for different classes of systems under certain assumptions concerning the parameters of the system and the available information.

In 1963 David Luenberger laid the foundations of the theory of observers for linear stationary control systems. Still now works appear which generalize or extend this theory to new classes of systems.

We shall briefly consider questions which arise in the theory of *asymptotic observers* which solve the observation problem in asymptotics when time tends to infinity, in contrast, say, to *finite observers* which solve a problem in finite time. In what follows we speak only about asymptotic observers.

The first problem consists in obtaining an answer to the question whether it is possible, in principle, to reconstruct (to construct an estimate) a full-phase vector for the given system using the available information. This problem is known as a *problem of observability of a dynamical system.*

The complete solution of this problem have been obtained for many kinds of dynamical control systems, including linear stationary multiply connected control systems which are described by equations of the form

$$\begin{cases} \dot{x} = Ax + Bu \\ y = Cx, \end{cases} \tag{1.3}$$

where $x \in \mathbb{R}^n$ is an unknown phase vector of the system, $u \in \mathbb{R}^m$, $y \in \mathbb{R}^l$ are the known input and output of the system respectively, A, B, and C are constant matrices of the corresponding dimensions.

The observability problem is also solved for linear nonstationary systems, i.e., for vector systems of form

$$\begin{cases} \dot{x} = A(t)x + B(t)u \\ y = C(t)x \end{cases} \tag{1.4}$$

under certain conditions imposed on the matrix coefficients $A(t)$, $B(t)$, and $C(t)$.

More complicated is the situation with nonlinear systems of the general form (1.1), (1.2). However, for many special cases this problem has been solved. For the systems which admit of the reconstruction of the phase vector from the available information (systems of this kind are said to be *observable*) a problem arises of obtaining an estimate $\tilde{x}(t)$ of the phase vector $x(t)$.

For solving this problem we traditionally use auxiliary dynamical systems which form the indicated estimate. In the general esse systems of this kind can be written as

$$\begin{cases} \dot{z} = g(z, u, y) \\ \tilde{x} = p(z, u, y). \end{cases} \tag{1.5}$$

Precisely these systems are called *observers.* Here the functions $g(\cdot)$ and $p(\cdot)$ are synthesized, and the dimension of the vector $z(t)$ is called the dimension of the observer. If the estimate $\tilde{x}(t)$ asymptotically converges to the phase vector of the system $x(t)$, then the observer is said to be *asymptotic* (if, in addition, the estimate $\|\tilde{x}(t) - x(t)\| \le C_0 \|\tilde{x}(0) - x(0)\| e^{-\gamma t}$ holds[1], where the constants $\gamma > 0$, $C_0 > 0$, then an observer of this kind is said to be *exponential*). For linear stationary fully determined systems (1.3) this problem has been completely solved.

However, for linear systems with uncertainty (systems with disturbances) of the form

$$\begin{cases} \dot{x} = Ax + Bu + D\xi \\ y = Cx, \end{cases} \tag{1.6}$$

[1] $\| \ \|$ is a norm in $\mathbb{R}^n$.

where $\xi \in \mathbb{R}^k$ is an unknown disturbance, the problem concerning the synthesis of asymptotic observers has not been completely solved. Papers still appear in which approaches are proposed to the solution of the indicated problem under different assumptions concerning the parameters of system (1.6) and the unknown perturbation ξ.

Still more complicated is the situation with the synthesis of observers in a nonlinear case, this problem has been solved only for certain classes of nonlinear dynamical systems.

Rather frequently, in the theory of automatic control, in addition to the stability of a closed-loop system, some additional requirements are presented to the properties of the regulator. In particular, it is often required that the dimension of the observer[2] (i.e., the dimension of the phase vector $z(t)$ of the dynamical system (1.5)) should be minimal. As a result, a problem appeared connected with the construction of a *minimal observer*, namely, an observer of the minimal dynamical order, i.e., of a minimal dimension.

For linear, stationary fully determined systems (1.3) this problem of estimation of the full-phase vector was completely solved in papers by Luenberger. At the same time, in order to solve control problems we often don't need know the whole phase vector of the system but may only use information about a certain functional of this vector, say, of the form

$$\sigma = h(x) \in \mathbb{R}^p, \tag{1.7}$$

where $h(\cdot)$ is a known sufficiently smooth function. In this case, we have a problem of constructing an estimate for this functional, or, in other words, a problem of constructing a *functional observer*. It states to reason that this problem has sense when the dimension of this observer is lower than the dimension of the observer which reconstructs the full-phase vector.

In the case of a linear stationary system without uncertainty and a linear functional

$$\sigma = Hx$$

this problem was considered in the monograph by O'Reilly [87], who proposed methods for constructing functional observers and obtained an upper estimate for the dimensions of these observers. However, the problem about a functional observer of minimal dimension was solved only recently.

In addition, of an individual interest is a problem concerning the synthesis of functional observers for linear and nonlinear stationary and nonstationary systems with uncertainty.

Similar problems, i.e., problems concerning observability, on the synthesis of an observer, on the construction of an observer under the conditions of uncertainty, on the synthesis of functional observers, on a minimal observer are also encountered in the case of discrete regulative systems, in particular for linear discrete control systems

[2] An observer is, as a rule, a part of the regulator.

which are described by the equations

$$\begin{cases} x_{k+1} = Ax_k + Bu_k \\ \quad y_k = Cx_k, \quad k = 0, 1, 2, \dots, \end{cases}$$

where, as before, $x \in \mathbb{R}^n$ is a phase vector, $u \in \mathbb{R}^m$ and $y \in \mathbb{R}^l$ are the input and the output of the system, respectively.

For linear stationary systems the majority of the results can be generalized from a continuous case to a discrete one although for the latter case there exist peculiarities and essential differences.

Chapter 2

Observability

2.1 Observability, identifiability, observability and identifiability criteria

Consider a problem of observability, i.e., a problem of possibility, in principle, of reconstruction of the phase vector of the system using the measurements of its output.

In the sequel, we consider a linear system of the form

$$\begin{cases} \dot{x} = A(t)x + B(t)u \\ y = C(t)x, \end{cases} \tag{2.1}$$

where $x \in \mathbb{R}^n$ is an unknown phase vector, $u \in \mathbb{R}^m$, $y \in \mathbb{R}^l$ are known input and output of the system, respectively.

We shall call the pair (t^*, x^*) the state of the system (at the time moment t^*) if $x^* = x(t^*)$. The solution of the system corresponding to the control $u(t)$ and the initial state (t_0, x_0) will be denoted by $x(t, t_0, x_0, u)$ and the output by $y(t, t_0, x_0, u)$ respectively, where $t \geq t_0$.

Two problems of reconstruction of the unknown vector $x(t)$ are distinguished.

An *observation problem* is the problem of estimation of the state of the system at the time moment t_0 from the known input and output of the system $u(t)$ and $y(t)$ for $t \geq t_0$, i.e., the problem of reconstruction of the initial value of the phase vector from the future measurements of the input and output.

A *problem of identifiability* is the problem of estimation of the state of the system at the time moment t^* from the data on the input and output for $t \leq t^*$, i.e., the problem of reconstruction of the phase vector at the time moment t^* from the measurements of the input and output at the past time.

These definitions were given in Kalman's papers.

Remark 2.1. Many authors do not distinguish between observability and identifiability combining these notions by the term observability. Sometimes an observable system is defined as a system in which past values of the output and input can be used in order to reconstruct the present state of the system. Somewhat above this problem was defined as a problem of identification.

In what follows, for simplicity, we shall denote as $x(t, t_0, x_0, u) = x(t)$ the solution of system (2.1) corresponding to the initial state $x(t_0) = x_0$ and the input $u(t)$. The corresponding output of the system will be denoted as $y(t, t_0, x_0, u) = y(t)$.

We introduce the notions of an observable and identifiable system (following [1]).

Definition 2.2. The linear system (2.1) is *observable at the time moment* t_0 if $y(t; t_0, x_0, 0) \equiv 0$ for $t \geq t_0$ if and only if $x_0 = 0$.

Definition 2.3. The linear system (2.1) is *identifiable at the time moment* t_0 if $y(t; t_0, x_0, 0) \equiv 0$ for $t \leq t_0$ if and only if $x_0 = 0$.

Note that if a system is identifiable (observable), then, at the nonzero initial state (t_0, x_0) (and the zero input $u \equiv 0$), the output of the system is not identically zero, i.e., the nonzero initial state generates a certain nontrivial reaction of the output.

Let us now consider linear stationary regularized systems of the form

$$\begin{cases} \dot{x} = Ax + Bu \\ y = Cx, \end{cases} \tag{2.2}$$

where $x \in \mathbb{R}^n$ is a phase vector, $u \in \mathbb{R}^m$ and $y \in \mathbb{R}^l$ are known input and output of the system, A, B, C are constant matrices of the corresponding dimensions. Since the observability and identifiability of system (2.2) are defined entirely by the matrices A and C, we speak about the observability (identifiability) of the pair $\{C, A\}$.

The following statement holds for the stationary linear system (2.2).

Theorem 2.4. *The stationary pair* $\{C, A\}$ *is observable if and only if it is identifiable.*

In the sequel we shall speak only about the observability of the pair $\{C, A\}$. The simple criterion of observability of the pair $\{C, A\}$ holds for linear stationary systems (2.2).

Theorem 2.5. *The stationary pair* $\{C, A\}$ *is observable if and only if the following rank condition is fulfilled*:

$$\operatorname{rank} \begin{pmatrix} C \\ CA \\ \vdots \\ CA^{n-1} \end{pmatrix} = n. \tag{2.3}$$

The matrix $N(C, A) = \begin{pmatrix} C \\ CA \\ \vdots \\ CA^{n-1} \end{pmatrix}$ is called a *matrix of observability* (Kalman's observability matrix).

Proof. Sufficiency. Since system (2.2) is stationary, we can set $t_0 = 0$. If the system is nonobservable, then there exists a vector $x_0 \neq 0$ such that

$$y(t, 0, x_0, 0) = Ce^{At} x_0 = 0$$

for all $t \geq 0$.

Successively differentiating $(n-1)$ times the output $y(t)$ by virtue of system (2.2) for $u(t) \equiv 0$, we obtain a system of equations

$$\begin{aligned} y(0) &= Cx_0 = 0 \\ y'(0) &= CAx_0 = 0 \\ &\vdots \\ y^{(n-1)}(0) &= CA^{n-1}x_0 = 0. \end{aligned}$$

Since $x_0 \neq 0$, this means that $\operatorname{rank} N(C, A) < n$. Consequently, if the matrix of observability $N(C, A)$ is of full rank, then system (2.2) is observable.

Necessity. Suppose that system (2.2) is observable. Let us show that $N(C, A)$ is a matrix of full rank.

Let $\operatorname{rank} N(C, A) < n$. Then there exists a vector $x_0 \neq 0$ such that

$$Cx_0 = 0, \quad CAx_0 = 0, \quad \ldots, \quad CA^{n-1}x_0 = 0. \tag{2.4}$$

By virtue of the Cayley–Hamilton theorem the matrices A^q for $q \geq n$ are expressed in terms of the matrices $I, A, \ldots, A^{n-1}$, and therefore it follows from (2.4) that $CA^q x_0 = 0$ for all $q \geq 0$.

Then the relation

$$y(t) = Ce^{At}x_0 = C\left[\sum_{i=0}^{\infty} \frac{A^i t^i}{i!}\right] x_0 = \sum_{i=0}^{\infty} \left[\frac{CA^i x_0}{i!} t^i\right] = 0$$

is also valid for the matrix exponent for all $t \geq 0$. Consequently, if $\operatorname{rank} N(C, A) < n$, then there exists a nonobservable state $(0, x_0)$, and this contradicts the assumption that the pair $\{C, A\}$ is observable. The theorem is proved. □

The rank condition (2.3) means that among the (nl) rows of the matrix $N(C, A) \in \mathbb{R}^{(nl)\times n}$ there are n linearly independent rows. It may turn out that the condition

$$\operatorname{rank} \begin{pmatrix} C \\ CA \\ \vdots \\ CA^{\nu-1} \end{pmatrix} = \operatorname{rank} N_\nu(C, A) = n \tag{2.5}$$

holds for a certain $\nu \leq n$. The minimal number ν for which condition (2.5) is fulfilled is called an *observability index* of the pair $\{C, A\}$ (of system (2.2)). Sometimes the matrix $N_\nu(C, A)$ is called an observability matrix.

The following statement is valid for the stationary system (2.2).

Lemma 2.6. *If* rank $N_p(C, A) = \text{rank}\, N_{p+1}(C, A)$, *then*

$$\text{rank}\, N_p(C, A) = \text{rank}\, N_{p+q}(C, A)$$

for all $q \geq 1$.

Proof. If

$$\text{rank}\, N_p(C, A) = \text{rank} \begin{pmatrix} C \\ CA \\ \vdots \\ CA^{p-1} \end{pmatrix} = \text{rank} \begin{pmatrix} C \\ CA \\ \vdots \\ CA^{p-1} \\ CA^p \end{pmatrix} = \text{rank}\, N_{p+1}(C, A),$$

then this means that the rows of the matrix CA^p can be linearly expressed in terms of the rows of the matrices $C, CA, \ldots, CA^{p-1}$. In that case, the rows of the matrix $CA^{p+1} = CA^pA$ can be linearly expressed in terms of the rows of the matrices $C, CA, \ldots, CA^{p-1}$. Consequently,

$$\text{rank}\, N_p(C, A) = \text{rank}\, N_{p+2}(C, A).$$

Continuing the discussion by induction, we get the statement of the lemma. The lemma is proved. □

Consequently, upon an increase of p, the rank of the matrices $N_p(C, A)$ either increases at every step or does not change beginning with a certain p^*. If rank $N_{p^*}(C, A) = n$, then the pair $\{C, A\}$ is observable and $p^* = \nu$. Now if rank $N_{p^*}(C, A) < n$, then the pair $\{C, A\}$ is nonobservable.

From Lemma 2.6 we obtain a simple corollary.

Corollary 2.7. *If the rank of the matrix C is maximal, i.e.,* rank $C = l$, *and the pair $\{C, A\}$ is observable, then*

$$\text{rank}\, N_{n-l+1}(C, A) = \text{rank} \begin{pmatrix} C \\ CA \\ \vdots \\ CA^{n-l} \end{pmatrix} = n.$$

Proof. By the definition rank$(C) = \text{rank}\, N_1(C, A) = l$. When we add rows of CA^p the rank of the matrices $N_p(C, A)$ increases at least by 1 until it reaches n. Consequently, the addition of rows of the matrices $CA, \ldots, CA^{n-l}$ increases rank $N_{n-l+1}(C, A)$ to n. The corollary is proved. □

Thus, if rank $C = l$, then the following estimate is valid for the observability index ν:

$$\nu \leq n - l + 1.$$

Let us consider the transformation of coordinates with the matrix P in the linear stationary system (2.2)

$$\bar{x} = Px,$$

where $\bar{x}$ are new coordinates.

Upon the indicated change the triple of matrices $\{C, A, B\}$ passes into a triple $\{\bar{C}, \bar{A}, \bar{B}\}$ which are connected with the initial matrices by the relations

$$\bar{C} = CP^{-1}, \quad \bar{A} = PAP^{-1}, \quad \bar{B} = PB.$$

Consequently, the observability matrix of the transformed system assumes the form

$$N(\bar{C}, \bar{A}) = \begin{pmatrix} CP^{-1} \\ CP^{-1}(PA)P^{-1} \\ \vdots \\ CP^{-1}(PAP^{-1})^{n-1} \end{pmatrix} = N(C, A)P^{-1}.$$

Since the transformation matrix P is nondegenerate, we have

$$\operatorname{rank} N(\bar{C}, \bar{A}) = \operatorname{rank} N(C, A),$$

i.e., the observability property is invariant to the change of coordinates. The same is, of course, true for $N_\nu(C, A)$.

For linear nonstationary systems with matrices $A(t)$, $C(t)$ in the case where $A(t)$ and $C(t)$ are functions differentiable a sufficient number of times, we can also define the observability matrix

$$N(t) = \begin{pmatrix} Q_1(t) \\ Q_2(t) \\ \vdots \\ Q_n(t) \end{pmatrix}, \tag{2.6}$$

where the matrices $Q_i(t) \in \mathbb{R}^{l \times n}$ are defined by the relations

$$Q_1(t) = C(t), \quad Q_{i+1}(t) = Q_i(t)A(t) + \dot{Q}_i(t), \quad i = 1, 2, \ldots, n-1.$$

The following definition is often used in literature.

Definition 2.8. System (2.1) (pair $\{C(t), A(t)\}$) is *uniformly (differentially) observable* if and only if the observability matrix $N(t) \in \mathbb{R}^{(nl) \times n}$ from (2.6) satisfies the rank condition

$$\operatorname{rank} N(t) = n, \quad t \geq t_0.$$

The problem of observability for linear systems is closely connected with the control problem. We shall briefly expose the main results following [1].

Consider a linear nonstationary system

$$\dot{x} = A(t)x + B(t)u. \tag{2.7}$$

Definition 2.9. The event (t_0, x_0) connected with the linear system (2.7) is said to be *controllable relative to the point* x_1 if there exists a time moment $t_1 \geq t_0$ and a control $u(t)$ defined on the interval $[t_0, t_1]$ which transforms the event (t_0, x_0) into an event (t_1, x_1).

For linear systems it is ordinary to consider the control relative to the origin, i.e., relative to $x_1 = 0$.

Definition 2.10. A linear system is said to be *controllable at the time moment* t_0 if every event (t_0, x), where t_0 is fixed and x is an arbitrary vector from $\mathbb{R}^n$, is controllable (relative to $x_1 = 0$).

A linear system is *controllable* (uniformly with respect to t_0) if it is controllable at any time moment t_0.

In the sequel we deal with controllability (*controllability of the pair* $\{A, B\}$) for the linear stationary system (2.2), the controllability criterion holding true.

Theorem 2.11. *The linear stationary system*

$$\dot{x} = Ax + Bu$$

is controllable (the pair $\{A, B\}$ is controllable) if and only if the rank condition

$$\operatorname{rank}(B, AB, \dots, A^{n-1}B) = n$$

is fulfilled.

The matrix $K(A, B) = (B, AB, \dots, A^{n-1}B) \in \mathbb{R}^{n\times(mn)}$ is called a *controllability matrix* (*Kalman's controllability matrix*).

The following statement similar to the statement of Lemma 2.6 is valid both for controllability matrices and for observability matrices.

Lemma 2.12. *For matrices $K_p(A, B) = (B, AB, \dots, A^{p-1}B)$, where $p = 0, 1, \dots,$ the rank of the matrices $K_p(A, B)$ monotonically increases to a certain p^* and for $p \geq p^*$ we have*

$$\operatorname{rank} K_p(A, B) = \operatorname{rank} K_{p^*}(A, B).$$

If the pair $\{A, B\}$ is controllable (i.e., system (2.2) is controllable), then the minimal number μ such that

$$\operatorname{rank} K_\mu(A, B) = n$$

is called a *controllability index*. If rank $B = m$, then rank $K(A, B) = (B, AB, \dots, A^{n-m}B) = n$, and the controllability index $\mu \le n - m + 1$.

Controllability and observability problems are dual [64]. Thus, for instance, if the stationary system

$$\begin{cases} \dot{x} = Ax + Bu \\ y = Cx \end{cases}$$

is controllable (observable), then *its dual*, i.e., a system of the form

$$\begin{cases} \dot{x}' = A^\top x' + C^\top u' \\ y' = B^\top x' \end{cases}$$

is observable (controllable).

Let us return now to an observability problem and consider a situation where the observability criterion is not fulfilled.

If rank $N(C, A) < n$, then system (2.2) is said to be *nonobservable* (*not completely observable*). Suppose that the condition

$$\operatorname{rank} N(C, A) = \xi$$

is fulfilled, where $0 < \xi < n$. For a not completely observable system there exists a nondegenerate transformation of coordinates [3, 63, 64] which reduces the system to the form

$$\begin{cases} \begin{cases} \dot{x}^1 = A_{11}x^1 + B_1 u \\ \dot{x}^2 = A_{21}x^1 + A_{22}x^2 + B_2 u \end{cases} \\ \quad y = C_1 x^1, \end{cases} \tag{2.8}$$

where $x^1 \in \mathbb{R}^\xi$, $x^2 \in \mathbb{R}^{n-\xi}$, A_{11}, A_{21}, A_{22}, B_1, B_2, and C_1 are matrices with constant coefficients of the corresponding dimensions. In this case, the pair $\{C_1, A_{11}\}$ is observable, x^1 is an observable part of the system, and x^2 is a nonobservable part of the system.

We have a similar result in the case where the criterion of controllability is not fulfilled. If rank $K(B, A) < n$, then system (2.2) is said to be *noncontrollable* (*not completely controllable*). Suppose that the condition

$$\operatorname{rank} K(B, A) = \eta, \quad 0 < \eta < n,$$

is fulfilled. Then, for a not completely controllable system there exists a nondegenerate transformation of coordinates [3, 64] which reduces the system to the form

$$\begin{cases} \begin{cases} \dot{x}^1 = A_{11}x^1 + A_{12}x^2 + B_1 u \\ \dot{x}^2 = A_{22}x^2 \end{cases} \\ \quad y = C_1 x^1 + C_2 x^2, \end{cases} \tag{2.9}$$

where $x^1 \in \mathbb{R}^\eta$, $x^2 \in \mathbb{R}^{n-\eta}$, A_{11}, A_{12}, A_{22}, B_1, C_1, and C_2 are constant matrices of the corresponding dimensions. In this case, the pair $\{A_{11}, B_1\}$ is controllable, i.e., x^1 is a controllable part of the system and x^2 is a noncontrollable one.

We say that system (2.2) *is in the general position* if it is controllable and observable (the *triple* $\{C, A, B\}$ *is in the general position* if the pair $\{C, A\}$ is observable and the pair $\{A, B\}$ is controllable).

If a system is not completely controllable and not completely observable, then a nondegenerate transformation of coordinates can reduce it to the form [3, 63, 64] called *Kalman's decomposition of the system*

$$\begin{cases} \begin{cases} \dot{x}^1 = A_{11}x^1 + A_{12}x^2 + A_{13}x^3 + B_1u \\ \dot{x}^2 = A_{22}x^2 + A_{24}x^4 + B_2u \\ \dot{x}^3 = A_{33}x^3 + A_{34}x^4 \\ \dot{x}^4 = A_{44}x^4 \end{cases} \\ \quad y = C_2x^2 + C_4x^4, \end{cases} \tag{2.10}$$

where $x^i \in \mathbb{R}^{n_i}$, A_{ij}, B_i, C_j are constant matrices of the corresponding dimensions. Here x^1 is a controllable but nonobservable part, x^2 is a controllable and observable part, x^3 is a noncontrollable and nonobservable part, x^4 is an observable but noncontrollable part.

In this case, $n_2 \le \min(\xi, \eta)$; $n_1 + n_2 = \eta$; $n_2 + n_4 = \xi$; $n_1 + n_2 + n_3 + n_4 = n$.

Let us consider a noncompletely observable system written in the canonical form (2.8). The following definition is valid.

Definition 2.13. A noncompletely observable system (2.2) is *reconstructible (detectable)* if the nonobservable coordinates of the system for $u \equiv 0$ and the identically zero observable part tend to zero as $t \to \infty$ (i.e., if A_{22} is a Hurwitz matrix[1] in the canonical representation (2.8)).

Definition 2.14. The not completely controllable system (2.2) is *stabilizable* if the noncontrollable coordinates tend to zero as $t \to \infty$ (i.e., if A_{22} is a Hurwitz matrix in the canonical representation (2.9)).

There exist a number of equivalent forms of criteria of controllability and observability. In some situations the algebraic criterion of observability of the linear stationary system (2.2) is convenient (in the Rosenbrock form formulated and proved in [85]).

---

[1] A *Hurwitz* matrix is a constant matrix whose characteristic polynomial satisfies the criterion of Hurwitz asymptotic stability.

Theorem 2.15 (Rosenbrock observability criterion). *The pair $\{C, A\}$ is observable if and only if the rank condition*

$$\operatorname{rank}\begin{pmatrix}\lambda I - A\\ C\end{pmatrix} = n, \quad \lambda \in \mathbb{C}, \tag{2.11}$$

is fulfilled.

Remark 2.16. Since the relation

$$\operatorname{rank}(\lambda I - A) = n$$

holds for all $\lambda \notin \operatorname{spec}\{A\}$, condition (2.11) should be verified only for $\lambda_i \in \operatorname{spec}\{A\}$, $i = 1, \dots, n$.

Proof. Necessity. Suppose that the pair $\{C, A\}$ is observable but there exists a number $\lambda^* \in \operatorname{spec}\{A\}$, $\lambda^* \in \mathbb{R}$ such that

$$\operatorname{rank}\begin{pmatrix}\lambda^* I - A\\ C\end{pmatrix} < n.$$

Then there exists a vector $x_0 \in \mathbb{R}^n$ such that

$$\begin{pmatrix}\lambda^* I - A\\ C\end{pmatrix} x_0 = 0, \quad x_0 \neq 0.$$

In this case, $x(t) = x_0 e^{\lambda^* t}$ is a solution of system (2.2) for $u(t) \equiv 0$ and the output of the system $y(t) = C x_0 e^{\lambda^* t} \equiv 0$. If $\lambda^* \in \mathbb{C}$, then $\bar{\lambda}^* \in \operatorname{spec}\{A\}$, $x_0 \in \mathbb{C}^n$ as well, and, in addition,

$$\begin{pmatrix}\bar{\lambda}^* I - A\\ C\end{pmatrix} \bar{x}_0 = 0,$$

and, consequently, $y(t) = C\bar{x}_0 e^{\bar{\lambda}^* t} \equiv 0$. Thus,

$$C(\gamma_1 x_0 e^{\lambda^* t} + \gamma_2 \bar{x}_0 e^{\bar{\lambda}^* t}) \equiv 0$$

for all γ_1 and $\gamma_2 \in \mathbb{C}$, and, hence, there exist numbers γ_1 and γ_2 such that the real function $\gamma_1 x_0 e^{\lambda^* t} + \gamma_2 \bar{x}_0 e^{\bar{\lambda}^* t}$ is nonzero and the output of the system

$$y(t) = C(\gamma_1 x_0 e^{\lambda^* t} + \gamma_2 \bar{x}_0 e^{\bar{\lambda}^* t}) \equiv 0.$$

Hence we have obtained a contradiction with the definition of observability of the system.

Sufficiency. Consider equations (2.2) for $u \equiv 0$ in the case where the output of the system $y(t) \equiv 0$

$$\begin{cases} \dot{x} = Ax \\ y = Cx. \end{cases}$$

We perform a Laplace transformation of this system under zero initial conditions. Denoting $X(s) = Lx(t)$, $Y(s) = Ly(t)$, we obtain an equation

$$\begin{cases} (sI - A)X = 0 \\ \qquad CX = 0. \end{cases}$$

Since the matrix $\binom{\lambda I - A}{C}$ is nondegenerate for all $\lambda \in \mathbb{C}$, it follows from the last system that $X(s) \equiv 0$. Making an inverse Laplace transformation, we find that $x(t) \equiv 0$ for $t \geq 0$, and, consequently, the identically zero output is associated only with an identically zero state vector. The theorem is proved. □

The Rosenbrock controllability criterion has a similar formulation, namely, the following theorem is valid.

Theorem 2.17 (Rosenbrock controllability criterion). *The pair $\{A, B\}$ is controllable if and only if the rank condition*

$$\operatorname{rank}(\lambda I - A, B) = n, \quad \lambda \in \mathbb{C}, \tag{2.12}$$

is fulfilled.

Remark 2.18. It is sufficient to verify condition (2.2) for $\lambda_i \in \operatorname{spec}\{A\}$, $i = 1, \ldots, n$.

The following statement is valid for a not completely observable system.

Theorem 2.19. *The not completely observable pair $\{C, A\}$ can be reconstructed if and only if the rank condition*

$$\operatorname{rank}\begin{pmatrix} \lambda I - A \\ C \end{pmatrix} = n, \quad \lambda \notin \mathbb{C}_-, \tag{2.13}$$

is fulfilled, where $\mathbb{C}_-$ is the left-hand open half-plane of the complex plane $\mathbb{C}$ (i.e., $\lambda \in \mathbb{C}_-$ if and only if $\operatorname{Re}\lambda < 0$).

Proof. Note that the rank of the Rosenbrock observability matrix

$$R(C, A) = \begin{pmatrix} \lambda I - A \\ C \end{pmatrix}$$

is invariant to the nondegenerate change of variables of the system. Indeed, upon the transition from the variables x to the variables $\bar{x} = Px$ the matrices of the system

$\{C, A\}$ pass to a pair $\{\bar{C}, \bar{A}\} = \{CP^{-1}, PAP^{-1}\}$. In this case, the Rosenbrock observability matrices for the old system and for the new one are related as

$$R(\bar{C}, \bar{A}) = \begin{pmatrix} \lambda I - PAP^{-1} \\ CP^{-1} \end{pmatrix} = \begin{pmatrix} P & 0 \\ 0 & I_l \end{pmatrix} \begin{pmatrix} \lambda I - A \\ C \end{pmatrix} P^{-1} = \begin{pmatrix} P & 0 \\ 0 & I_l \end{pmatrix} R(C, A) P^{-1},$$

where I_l is an identity matrix of order $l \times l$. By virtue of the nondegeneracy of the matrices appearing in the last product on the left and on the right, it follows that

$$\operatorname{rank} R(\bar{C}, \bar{A}) = \operatorname{rank} R(C, A)$$

for all $\lambda \in \mathbb{C}$, i.e., the lowering of the rank of these matrices occurs on the same values of λ and by the same number.

Therefore, in order to prove the theorem, it suffices to consider the system written in the canonical form (2.8). In this case, the matrices C and A have a block structure

$$A = \begin{pmatrix} A_{11} & 0 \\ A_{21} & A_{22} \end{pmatrix}, \quad C = (C_1, 0),$$

and, since the pair $\{C_1, A_{11}\}$ is observable,

$$\operatorname{rank} R(C_1, A_{11}) = \operatorname{rank} \begin{pmatrix} \lambda I_\nu - A_{11} \\ C_1 \end{pmatrix} = \nu, \quad \lambda \in \mathbb{C},$$

where ν is the observability index of the pair $\{C_1, A_{11}\}$. Let us write the Rosenbrock observability matrix for the pair $\{C, A\}$ in block form

$$R(C, A) = \begin{pmatrix} \lambda I_\nu - A_{11} & 0 \\ A_{21} & \lambda I_{n-\nu} - A_{22} \\ C_1 & 0 \end{pmatrix}.$$

Since $R(C_1, A_{11})$ has a full rank, the decrease of the rank of the matrix $R(C, A)$ occurs only when the rank of the matrix $(\lambda I_{n-\nu} - A_{22})$ decreases. This takes place on the eigenvalues of the matrix A_{22} which characterizes the dynamics of the nonobservable part of the system.

The pair $\{C, A\}$ is detectable if and only if A_{22} is a Hurwitz matrix. Consequently, for all $\lambda \notin \mathbb{C}_-$ the matrix $(\lambda I_{n-\nu} - A_{22})$, and, consequently, $R(C, A)$ has a full rank if and only if the pair $\{C, A\}$ is detectable. The theorem is proved. □

Remark 2.20. The output feedback of the system does not change the spectrum of its nonobservable subsystem. Indeed, let $L \in \mathbb{R}^{n \times l}$ be an arbitrary constant matrix and $A_L = A - LC$. Then, if the pair $\{C, A\}$ is observable, then the pair $\{C, A_L\}$ is also observable. If the pair $\{C, A\}$ is reconstructible, then the pair $\{C, A_L\}$ is also reconstructible, and the spectrum of the nonobservable parts is the same for both pairs. In order to prove this fact, it suffices to note that the relation

$$\operatorname{rank} \begin{pmatrix} sI - (A - LC) \\ C \end{pmatrix} = \operatorname{rank} \begin{pmatrix} sI - A \\ C \end{pmatrix}$$

holds for all $\lambda \in \mathbb{C}$.

A similar statement is valid for a not completely observable system as well.

Theorem 2.21. *The not completely controllable pair $\{A, B\}$ is stabilizable if and only if the rank condition*

$$\operatorname{rank}(\lambda I - A, B) = n, \quad \lambda \notin \mathbb{C}_-, \tag{2.14}$$

is fulfilled.

Remark 2.22. The state feedback does not change the spectrum of its noncontrollable subsystem. Indeed, let $K \in \mathbb{R}^{n\times m}$ be an arbitrary constant matrix and $A_K = A - BK$. Then, if the pair $\{A, B\}$ is controllable, then the pair $\{A_K, B\}$ is controllable as well. If the pair $\{A, B\}$ is stabilizable, then the pair $\{A_K, B\}$ is also stabilizable, the spectrum of the noncontrollable parts of both pairs being the same.

2.2 Transfer function and canonical forms

In the control theory of a very wide use is the concept of a *transfer function* which is defined as the operator $W(s)$ of the complex variable s which connects the Laplace transformations $Y(s)$ and $U(s)$ of the output and input of the system under zero initial conditions, i.e.,

$$Y(s) = W(s)U(s).$$

For the linear stationary system

$$\begin{cases} \dot{x} = Ax + Bu \\ y = Cx \end{cases}$$

the transfer function has the form

$$W(s) = C(sI - A)^{-1}B. \tag{2.15}$$

If $y \in \mathbb{R}^l$, $u \in \mathbb{R}^m$, then $W(s) \in \mathbb{C}^{l\times m}$, the elements of the matrix $W(s)$ being fractional-rational functions of the complex variable s. The representation

$$W(s) = \frac{C(\operatorname{adj}(sI - A))B}{\det(sI - A)} = \frac{(\beta_{ij}(s))}{\alpha(s)}, \tag{2.16}$$

holds for $W(s)$. Here $\operatorname{adj}(sI - A)$ is an algebraic complement of the corresponding matrix, $\alpha(s)$ is a characteristic polynomial of the matrix A, $(\beta_{ij}(s))$ is an $l \times m$ matrix of the polynomials of s, with $\deg(\beta_{ij}(s)) < \deg \alpha(s)$, $i = 1, \dots, l$, $j = 1, \dots, m$.

In the case of a scalar system, where $m = l = 1$, the transfer function $W(s)$ is a fractional-rational function

$$W(s) = \frac{\beta(s)}{\alpha(s)}.$$

Let us introduce the following notation for the coefficients of the polynomials $\beta(s)$ and $\alpha(s)$:

$$\begin{aligned}\alpha(s) &= \alpha_1 + \alpha_2 s + \cdots + \alpha_n s^{n-1} + s^n \\ \beta(s) &= \beta_1 + \beta_2 s + \cdots + \beta_n s^{n-1}\end{aligned} \tag{2.17}$$

(it is taken into account here that $\deg \alpha(s) = n$, $\deg \beta(s) < n$; in general, all leading coefficients of the polynomial $\beta(s)$, up to the coefficient with the number q, may be zero, i.e., $\beta_n = \beta_{n-1} = \cdots = \beta_{q+1} = 0$, $\beta_q \neq 0$. In this case, the number $r = n + 1 - q$ is called a *relative order* of the scalar system or, respectively, a relative order of the transfer function $W(s)$).

2.2.1 Canonical forms for scalar systems

We introduce the concept of the *(first) observable canonical representation of the system* for a scalar system when

$$A = \begin{pmatrix} 0 & 1 & 0 & \dots & 0 \\ 0 & 0 & 1 & \dots & 0 \\ \hdotsfor{5} \\ 0 & 0 & 0 & \dots & 1 \\ -\alpha_1 & -\alpha_2 & -\alpha_3 & \dots & -\alpha_n \end{pmatrix}, \quad B = \begin{pmatrix} CB \\ CAB \\ \vdots \\ CA^{n-1}B \end{pmatrix}, \tag{2.18}$$
$$C = (1, 0, \dots, 0).$$

Theorem 2.23. *The linear stationary system* (2.2) *for* $l = m = 1$ *can be reduced, by the nondegenerate transformation of coordinates, to the canonical form* (2.18) *if and only if the pair* $\{C, A\}$ *is observable.*

Proof. Necessity. If the system is reduced to form (2.18), then, in order to investigate the observability of the pair $\{C, A\}$, it suffices to find a Kalman observability matrix for the representation

$$N(C, A) = \begin{pmatrix} 1 & 0 & \dots & 0 \\ 0 & 1 & \dots & 0 \\ \hdotsfor{4} \\ 0 & 0 & \dots & 1 \end{pmatrix} = I, \quad \operatorname{rank} N(C, A) = n.$$

Sufficiency. If the pair $\{C, A\}$ is observable, then the vectors $C, CA, \dots, CA^{n-1}$ form a basis in $\mathbb{R}^n$. In this basis

$$C = (1, 0, \dots, 0) = e_1.$$

Let us find the rows of the matrix A in this basis:

$$\begin{aligned}
e_1 A &= CA = e_2 \\
e_2 A &= (CA)A = CA^2 = e_3 \\
&\vdots \\
e_{n-1} A &= (CA^{n-2})A = CA^{n-1} = e_n \\
e_n A &= (CA^{n-1})A = CA^n = C(-\alpha_1 I - \alpha_2 A - \alpha_n A^{n-1}) \\
&= -\alpha_1 e_1 - \alpha_2 e_2 - \cdots - \alpha_n e_n.
\end{aligned}$$

The last relation follows from the Cayley–Hamilton theorem.

Let us find the column B in the indicated basis. Note that the values of the variables $CA^i B$ are invariant under a change of variables. Indeed, upon the transition to the new basis with the matrix P, we have

$$\bar{C}\bar{A}^i\bar{B} = (CP^{-1})(PAP^{-1})^i(PB) = CA^i B.$$

Let $B = (b_1, \dots, b_n)^\top$. Then, taking into account the explicit representation for the matrices C and A in the new basis, we obtain

$$\begin{aligned}
CB &= b_1 \\
CAB &= b_2 \\
&\vdots \\
CA^{n-1}B &= b_n.
\end{aligned}$$

The theorem is proved. □

For the observable pair $\{C, A\}$ we have an alternative canonical representation (in the sequel we call it the *second observable representation*)

$$A = \begin{pmatrix} 0 & 0 & \dots & 0 & -\alpha_1 \\ 1 & 0 & \dots & 0 & -\alpha_2 \\ \dots & \dots & \dots & \dots & \dots \\ 0 & 0 & \dots & 1 & -\alpha_n \end{pmatrix}, \quad C = (0, \dots, 0, 1), \quad B = \begin{pmatrix} \beta_1 \\ \vdots \\ \beta_n \end{pmatrix}, \tag{2.19}$$

where α_i $(i = 1, \dots, n)$ and β_j $(j = 1, \dots, n)$ are coefficients of the polynomials from (2.17).

Theorem 2.24. *The linear stationary system* (2.2) *for* $l = m = 1$ *can be reduced, by a nondegenerate transformation of coordinates, to the canonical form* (2.19) *if and only if the pair* $\{C, A\}$ *is observable.*

Proof. Necessity. If the system is reduced to form (2.19), then, in order to investigate the observability, it suffices to investigate the pair $\{C, A\}$ given in this form. By a direct verification we find that

$$N(C, A) = \begin{pmatrix} 0 & \dots & 0 & 0 & 1 \\ 0 & \dots & 0 & 1 & * \\ 0 & \dots & 1 & * & * \\ \dots & \dots & \dots & \dots & \dots \\ 1 & \dots & * & * & * \end{pmatrix},$$

where $*$ are possibly nonzero elements dependent on the coefficients α_i. Since $\operatorname{rank} N(C, A) = n$, the pair $\{C, A\}$ is observable.

Sufficiency. Let the pair $\{C, A\}$ be observable. Then the rows $C, CA, \dots, CA^{n-1}$ form a basis in the space $\mathbb{R}^n$. In that case, the vectors

$$\begin{aligned} e_n &= C \\ e_{n-1} &= CA + \alpha_n C \\ &\vdots \\ e_2 &= CA^{n-2} + \alpha_n CA^{n-3} + \alpha_{n-1} CA^{n-4} + \dots + \alpha_3 C \\ e_1 &= CA^{n-1} + \alpha_n CA^{n-2} + \alpha_{n-1} CA^{n-3} + \dots + \alpha_2 C \end{aligned}$$

also form a basis. Indeed, the matrices of the direct and the inverse transfer from the basis $C, CA, \dots, CA^{n-1}$ to the basis $e_1, \dots, e_n$ are nondegenerate and have the form

$$P = \begin{pmatrix} 0 & 0 & \dots & 0 & 1 \\ 0 & 0 & \dots & 1 & \alpha_n \\ \dots & \dots & \dots & \dots & \dots \\ 0 & 1 & \dots & \alpha_4 & \alpha_3 \\ 1 & \alpha_n & \dots & \alpha_3 & \alpha_2 \end{pmatrix}, \quad P^{-1} = \begin{pmatrix} -\alpha_2 & -\alpha_3 & \dots & -\alpha_n & 1 \\ -\alpha_3 & -\alpha_4 & \dots & 1 & 0 \\ \dots & \dots & \dots & \dots & \dots \\ -\alpha_n & 1 & \dots & 0 & 0 \\ 1 & 0 & \dots & 0 & 0 \end{pmatrix}.$$

In the new basis $\{e_i\}$ $(i = 1, \dots, n)$ the vector C obviously has the required form

$$C = (0, \dots, 0, 1).$$

Let us find the representation of the matrix A in this basis. The first row of A has the form

$$\begin{aligned} e_1 A &= CA^n + \alpha_n CA^{n-1} + \dots + \alpha_2 CA \\ &= (CA^n + \alpha_n CA^{n-1} + \dots + \alpha_2 CA + \alpha_1 C) - \alpha_1 C. \end{aligned}$$

According to the Cayley–Hamilton theorem

$$C(A^n + \alpha_n A^{n-1} + \dots + \alpha_1) = 0,$$

and therefore

$$e_1 A = -\alpha_1 C = -\alpha_1 e_n.$$

Calculating the other rows A_i of the matrix A, we obtain

$$\begin{aligned} e_2 A &= CA^{n-1} + \alpha_n CA^{n-2} + \dots + \alpha_3 CA = e_1 - \alpha_2 e_n \\ &\vdots \\ e_{n-1} A &= CA^2 + \alpha_n CA = e_{n-2} - \alpha_{n-1} e_n \\ e_n A &= CA = e_{n-1} - \alpha_n e_n. \end{aligned}$$

Thus, the matrix A in the indicated basis also has the required form

$$A = \begin{pmatrix} 0 & 0 & \dots & 0 & -\alpha_1 \\ 1 & 0 & \dots & 0 & -\alpha_2 \\ \dots & \dots & \dots & \dots & \dots \\ 0 & 0 & \dots & 1 & -\alpha_n \end{pmatrix}.$$

Suppose that $B = (b_1, \dots, b_n)^\top$ in this basis. Let us find the transfer function of the system taking into account the explicit expression for A and C:

$$W(s) = C(sI - A)^{-1} B = \frac{C \operatorname{adj}(sI - A) B}{\det(sI - A)}.$$

The matrix A is a companion matrix for the polynomial $\alpha(s) = \alpha_1 + \alpha_2 s + \dots + \alpha_n s^{n-1} + s^n$, and therefore

$$\det(sI - A) = \alpha(s).$$

Since $C = (0, \dots, 0, 1)$, it follows that, in order to find the numerator of the transfer function, it suffices to find the last row of the matrix $\operatorname{adj}(sI - A)$ (i.e., $[\operatorname{adj}(sI - A)]_n$)

$$[\operatorname{adj}(sI - A)]_n = \left[\operatorname{adj} \begin{pmatrix} s & 0 & \dots & 0 & \alpha_1 \\ -1 & s & \dots & 0 & \alpha_2 \\ \dots & \dots & \dots & \dots & \dots \\ 0 & 0 & \dots & -1 & (s + \alpha_n) \end{pmatrix} \right]_n = (1, s, s^2, \dots, s^{n-1}).$$

Therefore the numerator of the transfer function

$$\beta(s) = (1, s, s^2, \dots, s^{n-1}) \begin{pmatrix} b_1 \\ \vdots \\ b_n \end{pmatrix} = b_1 + b_2 s + \dots + b_n s^{n-1}.$$

Taking into account the notation for the coefficients of the polynomial $\beta(s)$ from (2.17), we obtain

$$b_i = \beta_i, \quad B = (\beta_1, \dots, \beta_n)^\top.$$

The theorem is proved. □

Similar canonical forms are valid for a controllable system as well. A *controllable canonical representation* is the representation of system (2.2), where

$$A = \begin{pmatrix} 0 & 1 & 0 & \dots & 0 \\ 0 & 0 & 1 & \dots & 0 \\ \dots & \dots & \dots & \dots & \dots \\ 0 & 0 & 0 & \dots & 1 \\ -\alpha_1 & -\alpha_2 & -\alpha_3 & \dots & -\alpha_n \end{pmatrix}, \quad B = \begin{pmatrix} 0 \\ \vdots \\ 0 \\ 1 \end{pmatrix}, \quad C = (\beta_1, \dots, \beta_n). \tag{2.20}$$

The following statement holds true.

Theorem 2.25. *The linear stationary system* (2.2) *for* $l = m = 1$ *can be reduced, by a nondegenerate transformation of coordinates, to form* (2.20) *if and only if the pair* $\{A, B\}$ *is controllable.*

We omit the proof of the theorem since it is similar to the proof of Theorem 2.24.

2.2.2 Canonical forms for vector systems

We shall describe now canonical forms for vector systems, i.e., for the case $l > 1$ ($m > 1$). Suppose that, as before, the pair $\{C, A\}$ is observable. Then rank $N(C, A) = n$ and, among the rows of the observability matrix $N(C, A)$, we can choose n basis rows.

We denote the rows of the matrix C as C_i, $i = 1, \dots, l$, and then we shall choose the basis rows among the rows

$$\{C_1, \dots, C_l; C_1 A, \dots, C_l A; \dots; C_1 A^{n-1}, \dots, C_l A^{n-1}\}.$$

Here are two techniques of constructing a canonical basis.

Technique 1. We shall sort out the rows $C_1, C_1 A, \dots, C_1 A^{\nu_1 - 1}$ until the row $C_1 A^{\nu_1}$ will be expressed by the preceding rows. If $\nu_1 = n$, then the system is observable in terms of the output $y_1 = C_1 x$ (i.e., the pair $\{C_1; A\}$ is observable) and the problem reduces to the construction of a canonical form for a system with a scalar output.

If $\nu_1 < n$, then we successively add the rows $C_2, C_2 A, \dots, C_2 A^{\nu_2 - 1}$ until the row $C_2 A^{\nu_2}$ will be expressed in terms of the rows $\{C_1, C_1 A, \dots, C_1 A^{\nu_1 - 1}, C_2, \dots, C_2 A^{\nu_2 - 1}\}$.

If $\nu_1 + \nu_2 < n$, then we successively add the rows $C_3, C_3 A, \dots, C_3 A^{\nu_3 - 1}$, and so on.

As a result we obtain a system of n linearly independent vector-rows

$$\{C_1, C_1 A, \dots, C_1 A^{\nu_1 - 1}; C_2, \dots, C_2 A^{\nu_2 - 1}; \dots; C_k, \dots, C_k A^{\nu_k - 1}\}, \tag{2.21}$$

with $\nu_1 + \nu_2 + \dots + \nu_k = n$, $1 \le k \le l$. If $k < l$, then this means that the system is observable in terms of the output $\tilde{y} = (y_1, \dots, y_k)$ (i.e., the pair $\{\tilde{C}; \tilde{A}\}$ is observable, where $\tilde{C} = \begin{pmatrix} C_1 \\ \vdots \\ C_k \end{pmatrix} \in \mathbb{R}^{k \times n}$).

We take a set of vectors (2.21) as a new basis. Note that the relation $e_j A = e_{j+1}$ holds for the basis vectors e_j in the case where $j \neq \nu_1,\ \nu_1+\nu_2,\ \nu_1+\nu_2+\nu_3, \dots,$ $\nu_1+\dots+\nu_k = n$.

Now if $j = \nu_1,\ \nu_1+\nu_2, \dots, \nu_1+\dots+\nu_k = n$, then $e_j A = C_i A^{\nu_i}$, and, consequently, $e_j A$ is expressed in terms of the preceding basis vectors. Thus, in the new basis the matrices A and C have the structure

$$C = \begin{pmatrix} \bar{C}_1 & 0 & 0 & \dots & 0 \\ 0 & \bar{C}_2 & 0 & \dots & 0 \\ \dots & \dots & \dots & \dots & \dots \\ 0 & 0 & 0 & \dots & \bar{C}_k \\ \bar{C}^1_{k+1} & \bar{C}^2_{k+1} & \bar{C}^3_{k+1} & \dots & \bar{C}^k_{k+1} \end{pmatrix}. \tag{2.22}$$

Here $\bar{C}_i = (1, 0, \dots, 0) \in \mathbb{R}^{1 \times \nu_i}$, $\bar{C}^i_{k+1} \in \mathbb{R}^{(l-k) \times \nu_i}$ are some matrices;

$$A = \begin{pmatrix} A_{11} & 0 & \dots & 0 \\ A_{21} & A_{22} & \dots & 0 \\ \vdots & \vdots & \ddots & \vdots \\ A_{k1} & A_{k2} & \dots & A_{kk} \end{pmatrix}, \tag{2.23}$$

$$A_{ii} = \begin{pmatrix} 0 & 1 & 0 & \dots & 0 \\ 0 & 0 & 1 & \dots & 0 \\ \dots & \dots & \dots & \dots & \dots \\ 0 & 0 & 0 & \dots & 1 \\ * & * & * & \dots & * \end{pmatrix} \in \mathbb{R}^{\nu_i \times \nu_i}, \quad A_{ij} = \begin{pmatrix} 0 & 0 & \dots & 0 \\ \dots & \dots & \dots & \dots \\ 0 & 0 & \dots & 0 \\ * & * & \dots & * \end{pmatrix} \in \mathbb{R}^{\nu_j \times \nu_i}, \ j > i,$$

where $*$ are possibly nonzero elements of the matrices A_{ii} and A_{ij}.

Since in the indicated basis the matrix A has a block-triangular form, its characteristic polynomial is equal to the product of the characteristic polynomials of the diagonal matrices A_{ii} which, in turn, have the form of the companion matrices for some polynomials.

Also note that each one of the pairs $\{\bar{C}_i, A_{ii}\}$ has the first canonical form of observability. In addition, the system is observable as concerns the first k outputs, and therefore, without loss of observability, the last $(l-k)$ rows of the matrix C can be deleted.

The canonical form (2.22), (2.23) is called the *first canonical form of observability* for vector systems. Note that this form is not uniquely defined since we can begin the process of sorting out the vectors not from the vector C_1 but from any C_i. In this case, the number of cells and their size may depend on the order of this sorting out.

Here is the technique of constructing an alternative canonical basis for system (2.2).

Technique 2. We begin with choosing the rows $\{C_1, \dots, C_l\}$ from the rows of the observability matrix. If the matrix C is of full rank (and this is precisely supposed), then these rows are linearly independent. We successively add the rows $C_1 A$,

$C_2A, \dots, C_lA$ to this set so that the set will consist of linearly independent vectors. If C_iA is expressed in terms of the preceding vectors, then we do not include this vector into the set and pass to $C_{i+1}A$. Then we sort out the vectors C_iA^2, then C_iA^3, and so on, up to $C_iA^{\nu-1}$, where ν is the observability index of the pair $\{C, A\}$. Note that if C_iA^j is expressed via the preceding vectors, then, for $q > j$, C_iA^q is also expressed via these vectors. Therefore, as a result of the indicated procedure (after the renumbering) we obtain a set of basis vectors

$$\{C_1, C_1A, \dots, C_1A^{\nu_1-1}; C_2, C_2A, \dots, C_2A^{\nu_2-1}; \dots; C_l, C_lA, \dots, C_lA^{\nu_l-1}\},$$

where $\nu_i \geq 1$, $i = 1, \dots, l$; $\nu_1 + \nu_2 + \dots + \nu_l = n$, and also $\max \nu_i = \nu$.

Indeed, if $\max \nu_i < \nu$, then the basis set of vectors can be found among the rows of the matrix

$$\begin{pmatrix} C \\ CA \\ \vdots \\ CA^{(\max \nu_i)-1} \end{pmatrix} = N_{\max \nu_i}(C, A).$$

However, this means that $\operatorname{rank} N_{\max \nu_i}(C, A) = n$ and ν is not a minimal value for which the rank condition

$$\operatorname{rank} N_\nu(C, A) = n$$

is fulfilled, and this contradicts the definition of the observability index.

In the indicated basis the matrix C has a block-diagonal structure

$$C = \begin{pmatrix} \bar{C}_1 & 0 & \dots & 0 \\ 0 & \bar{C}_2 & \dots & 0 \\ \dots & \dots & \dots & \dots \\ 0 & 0 & \dots & \bar{C}_l \end{pmatrix}, \qquad \bar{C}_i = (1, 0, \dots, 0) \in \mathbb{R}^{\nu_i}. \tag{2.24}$$

Let us find the matrix A in this basis, for which purpose we note that

$$e_iA = e_{i+1}$$

if $i \neq \nu_1, \nu_1 + \nu_2, \dots, \nu_1 + \dots + \nu_l = n$. Now if $i = \nu_1, \nu_1 + \nu_2, \dots, \nu_1 + \dots + \nu_l = n$, then the vector e_iA is expressed via all basis vectors. Thus, the matrix A has the form

$$A = \begin{pmatrix} A_{11} & A_{12} & \dots & A_{1l} \\ A_{21} & A_{22} & \dots & A_{2l} \\ \dots & \dots & \dots & \dots \\ A_{l1} & A_{l2} & \dots & A_{ll} \end{pmatrix}, \tag{2.25}$$

where the diagonal elements have the form of the companion matrix for a certain polynomial

$$A_{ii} = \begin{pmatrix} 0 & 1 & 0 & \dots & 0 \\ 0 & 0 & 1 & \dots & 0 \\ \dots & \dots & \dots & \dots & \dots \\ 0 & 0 & 0 & \dots & 1 \\ * & * & * & \dots & * \end{pmatrix}, \qquad A_{ii} \in \mathbb{R}^{\nu_i \times \nu_i}. \tag{2.26}$$

Here $*$ are possibly nonzero elements. The off-diagonal matrices have the form

$$A_{ij} = \begin{pmatrix} 0 & 0 & \dots & 0 \\ 0 & 0 & \dots & 0 \\ \dots & \dots & \dots & \dots \\ 0 & 0 & \dots & 0 \\ * & * & \dots & * \end{pmatrix}, \quad A_{ij} \in \mathbb{R}^{\nu_i \times \nu_j}. \tag{2.27}$$

Note that the size of the maximal diagonal block is equal to the observability index of the pair $\{C, A\}$. We call the canonical form (2.24)–(2.27) the *second canonical form of observability* for vector systems.

Now we shall describe one more convenient canonical form of observability following Luenberger's work [78]. For this purpose, from the observability matrix we again choose rows as was proposed in Technique 2 and compose a matrix

$$V = \begin{pmatrix} C_1 \\ C_1 A \\ \vdots \\ C_1 A^{\nu_1 - 1} \\ --- \\ C_2 \\ \vdots \\ C_2 A^{\nu_2 - 1} \\ --- \\ \vdots \\ C_l A^{\nu_l - 1} \end{pmatrix}.$$

We find the matrix V^{-1} and denote by $g_i \in \mathbb{R}^n$ the column of the matrix V^{-1} with the number $\sigma_i = \nu_1 + \dots + \nu_i$ $(i = 1, \dots, l)$. The matrix of transition to the *Luenberger canonical basis* has the form

$$M = \left(g_1, Ag_1, \dots, A^{\nu_1 - 1} g_1, g_2, \dots, A^{\nu_2 - 1} g_2, \dots, A^{\nu_l - 1} g_l\right).$$

In the new basis $\bar{x} = M^{-1} x$ the matrices $\bar{A} = M^{-1} A M$ and $\bar{C} = CM$ have the form

$$\bar{A} = \begin{pmatrix} A_{11} & A_{12} & \dots & A_{1l} \\ A_{21} & A_{22} & \dots & A_{2l} \\ \dots & \dots & \dots & \dots \\ A_{l1} & A_{l2} & \dots & A_{ll} \end{pmatrix}, \tag{2.28}$$

where the diagonal matrices have the form of a companion matrix

$$A_{ii} = \begin{pmatrix} 0 & 0 & \dots & 0 & * \\ 1 & 0 & \dots & 0 & * \\ 0 & 1 & \dots & 0 & * \\ \dots & \dots & \dots & \dots & \dots \\ 0 & 0 & \dots & 1 & * \end{pmatrix}, \quad A_{ii} \in \mathbb{R}^{\nu_i \times \nu_i}, \tag{2.29}$$

and the off-diagonal blocks have the form

$$A_{ij} = \begin{pmatrix} 0 & \dots & 0 & * \\ 0 & \dots & 0 & * \\ \dots & \dots & \dots & \dots \\ 0 & \dots & 0 & * \end{pmatrix}, \quad A_{ij} \in \mathbb{R}^{\nu_i \times \nu_j}. \tag{2.30}$$

In the new basis the matrix C has the form

$$\bar{C} = \begin{pmatrix} \bar{C}_1 & 0 & 0 & \dots & 0 \\ \bar{C}_{21} & \bar{C}_2 & 0 & \dots & 0 \\ \bar{C}_{31} & \bar{C}_{32} & \bar{C}_3 & \dots & 0 \\ \dots & \dots & \dots & \dots & \dots \\ \bar{C}_{l1} & \bar{C}_{l2} & \bar{C}_{l3} & \dots & \bar{C}_l \end{pmatrix}, \tag{2.31}$$

where

$$\bar{C}_i \in \mathbb{R}^{1\times\nu_i}, \quad \bar{C}_i = (0,\dots,0,1), \quad \bar{C}_{ji} \in \mathbb{R}^{1\times\nu_i}, \quad \bar{C}_{ji} = (0,\dots,0,*).$$

Since the rows $\bar{C}_{ji}$ $(j = i+1,\dots,l)$ can be linearly expressed in terms of $\bar{C}_i$, the matrix $\bar{C}$ can be reduced to the form

$$\bar{C} = \begin{pmatrix} \bar{C}_1 & 0 & \dots & 0 \\ 0 & \bar{C}_2 & \dots & 0 \\ \dots & \dots & \dots & \dots \\ 0 & 0 & \dots & \bar{C}_l \end{pmatrix} \tag{2.32}$$

by a nondegenerate change of the output.

In the indicated basis the system is decomposed into l subsystems of order ν_i with phase vectors $x_i \in \mathbb{R}^{\nu_i}$. In this case, the components of the output y_i (after the transformation of the matrix $\bar{C}$ to a block-diagonal form (2.32)) correspond to the last coordinates of the vectors x_i. It follows from the explicit representation of the off-diagonal matrices

$$A_{ij} = \begin{pmatrix} 0 & \dots & 0 & * \\ 0 & \dots & 0 & * \\ \dots & \dots & \dots & \dots \\ 0 & \dots & 0 & * \end{pmatrix} = (0,\dots,0,\bar{a}_{ij}), \quad \bar{a}_{ij} \in \mathbb{R}^{\nu_i \times l},$$

that

$$A_{ij}x_j = \bar{a}_{ij}y_j.$$

Thus, in the *Luenberger canonical basis* the system can be written as

$$\begin{cases} \dot{x}_i = A_{ii}x_i + \sum\limits_{j=1,\, j\neq i}^{l} \bar{a}_{ij}y_j + B_i u, \quad i = 1,\dots,l \\ y_i = \bar{C}_i x_i, \end{cases} \tag{2.33}$$

i.e., the system decomposes into l subsystems of order ν_i each, the connection between which is realized via the components y_i of the output vector. The dimension of the maximal subsystem is exactly equal to the observability index of the pair $\{C, A\}$. Each of the pairs $\{\bar{C}_i, A_{ii}\}$ is observable, moreover, it is defined in the second canonical form of observability for scalar systems.

Note that in contrast to the first two observable canonical forms the transition to the Luenberger form is realized not only by the change of coordinates but also by the transformation of the outputs. Similar canonical forms can be indicated for controllable systems. We omit the details.

2.3 Canonical representation with the isolation of zero dynamics

Let us consider one more canonical representation of the system of general position (2.2) in which the *zero dynamics* of the system is isolated.

We assume, as before, that the system is in the general position, i.e., the pair $\{A, B\}$ is controllable and the pair $\{C, A\}$ is observable. In addition, we shall consider square systems, i.e., systems in which the dimensions of the input m and output l coincide (i.e., $u, y \in \mathbb{R}^l$).

2.3.1 Zero dynamics of scalar systems

We begin with considering the case of a scalar system, i.e., where $l = 1$. The scalar system (2.2) of the general position can be reduced, by means of a nondegenerate transformation of coordinates, to the canonical form of controllability:

$$\begin{cases} \begin{cases} \dot{x}_1 = x_2 \\ \quad\vdots \\ \dot{x}_{n-1} = x_n \\ \dot{x}_n = -\alpha_1 x_1 - \cdots - \alpha_n x_n + u \end{cases} \\ y = \beta_1 x_1 + \cdots + \beta_n x_n. \end{cases} \tag{2.34}$$

By tradition, we understand the zero dynamics of system (2.34) as a motion in the system belonging entirely to the manifold $y = Cx = 0$. For linear stationary systems the zero dynamics is also described by a system of linear stationary equations. Therefore we can define the characteristic polynomial of this system which we call in the sequel as a *characteristic polynomial of zero dynamics.*

For square systems the characteristic polynomial of zero dynamics is the determinant of the Rosenbrock matrix [86]

$$\beta(s) = \det R(s) = \det \left(\begin{array}{c|c} sI - A & -B \\ \hline C & 0 \end{array} \right). \tag{2.35}$$

The most simple is the definition of the characteristic polynomial of zero dynamics for scalar systems. In this case, the transfer function

$$W(s) = C(sI - A)^{-1}B = \frac{\beta(s)}{\alpha(s)}, \tag{2.36}$$

is defined, where $\alpha(s)$ and $\beta(s)$ are polynomials of s, $\deg\alpha(s) = n$, $\deg\beta(s) < n$. The polynomial $\alpha(s)$ is a characteristic polynomial of the matrix A and the polynomial $\beta(s)$ is a characteristic polynomial of zero dynamics. For scalar systems of the general position the *relative order* of the system is a number r such that the conditions

$$CB = 0, \quad CAB = 0, \quad \dots, \quad CA^{r-2}B = 0, \quad CA^{r-1}B \neq 0$$

are fulfilled, with $\deg\beta(s) = n - r$. It follows from the definition of the relative order that the first $(r-1)$ time derivatives of the output $y(t)$ do not depend explicitly on the input $u(t)$ but $y^{(r)}(t)$ depends explicitly on $u(t)$, to be more precise,

$$y^{(r)} = CA^r x + CA^{r-1}Bu.$$

It should also be noted that in the canonical form (2.34) α_i and β_j are coefficients of the polynomials $\alpha(s)$ and $\beta(s)$ respectively, i.e., the representations

$$\alpha(s) = s^n + \alpha_n s^{n-1} + \dots + \alpha_1$$
$$\beta(s) = \beta_n s^{n-1} + \beta_{n-1}s^{n-2} + \dots + \beta_1$$

are valid, where the leading coefficients β_j may be zero.

Let the relative order of the system be equal to r. Without loss of generality we can assume that $CA^{r-1}B = 1$ (this can always be achieved by the normalization of the output $y(t)$). Then

$$\beta(s) = s^{n-r} + \beta_{n-r}s^{n-r-1} + \dots + \beta_1,$$

where $\beta_{n-r+1} = CA^{r-1}B = 1$. In this case, in order to reduce system (2.34) to a canonical representation with the isolation of zero dynamics, we must pass to coordinates

$$x' = \begin{cases} x_1 \\ \vdots \\ x_{n-r}, \end{cases} \qquad y' = \begin{cases} y_1 = y = Cx \\ y_2 = \dot{y} = CAx \\ \vdots \\ y_r = y^{(r-1)} = CA^{r-1}x. \end{cases}$$

The matrix of transition from the coordinates x to coordinates $\binom{x'}{y'}$ has the form

$$P = \left(\begin{array}{cccc|cccc} 1 & 0 & \dots & 0 & 0 & 0 & \dots & 0 \\ \vdots & & \ddots & & \vdots & & & \vdots \\ 0 & 0 & & 1 & 0 & 0 & \dots & 0 \\ \hline \beta_1 & \beta_2 & \dots & \beta_{n-r} & 1 & 0 & \dots & 0 \\ 0 & \beta_1 & \dots & \beta_{n-r-1} & \beta_{n-r} & 1 & \dots & 0 \\ & & & & & \ddots & & \\ 0 & 0 & \dots & \dots & \dots & \dots & \dots & 1 \end{array}\right), \quad \det P = 1.$$

In the new coordinates the equations of the system assume the form

$$\begin{cases} \begin{cases} \dot{x}_1 = x_2 \\ \quad\vdots \\ \dot{x}_{n-r-1} = x_{n-r} \\ \dot{x}_{n-r} = -\beta_1 x_1 - \beta_2 x_2 - \dots - \beta_{n-r} x_{n-r} + y \end{cases} \\ \begin{cases} \dot{y}_1 = y_2 \\ \quad\vdots \\ \dot{y}_{r-1} = y_r \\ \dot{y}_r = -\sum_{i=1}^{n-r} \eta_i x_i - \sum_{j=1}^{r} \gamma_j y_j + u \end{cases} \\ y = y_1. \end{cases} \tag{2.37}$$

We can also write system (2.37) in compact form

$$\begin{cases} \dot{x}' = A_{11}x' + A_{12}y \\ \dot{y}_1 = y_2 \\ \quad\vdots \\ \dot{y}_{r-1} = y_r \\ \dot{y}_r = -\overline{\eta}x' - \overline{\gamma}y' + u, \end{cases} \tag{2.37$'$}$$

where the matrix A_{11} has the form of a companion matrix of the polynomial $\beta(s)$, $\overline{\eta} = (\eta_1, \dots, \eta_{n-r})$, $\overline{\gamma} = (\gamma_1, \dots, \gamma_r)$, $A_{12} = (0, \dots, 0, 1)^\top$.

The convenience of representation of the system in form (2.37′) is in the simplicity of obtaining an equation of zero dynamics of the system in the form

$$\dot{x}' = A_{11}x',$$

which is defined by the first $(n - r)$-dimensional part of the system whereas the other coordinates are derivatives of the output.

The following theorem is valid for the coefficients η_i and γ_j.

Theorem 2.26. *Suppose that we are given a system of general position* (2.2) *with scalar input and output, the transfer function of the system* $W(s) = C(sI - A)^{-1}B = \frac{\beta(s)}{\alpha(s)}$, *where* $\alpha(s)$ *and* $\beta(s)$ *are relatively prime polynomials* $\deg \alpha(s) = n$, $\deg \beta(s) = m$, *the leading coefficients of the polynomials are equal to* 1, *the polynomials* $\varphi(s)$ *and* $\psi(s)$ *are defined, the quotient and the remainder of the division of* $\alpha(s)$ *by* $\beta(s)$

$$\frac{\alpha(s)}{\beta(s)} = \varphi(s) + \frac{\psi(s)}{\beta(s)}, \quad \deg \varphi(s) = n - m = r, \quad \deg \psi(s) < m = n - r.$$

Then η_i *and* γ_i *from the canonical representation* (2.37) *are coefficients of the polynomials* $\psi(s)$ *and* $\varphi(s)$ *respectively, i.e.,*

$$\begin{aligned} \varphi(s) &= s^r + \gamma_r s^{r-1} + \cdots + \gamma_1 \\ \psi(s) &= s^{n-r-1}\eta_{n-r} + \cdots + \eta_1. \end{aligned} \tag{2.38}$$

Proof. Let the polynomials $\varphi(s)$ and $\psi(s)$ have structure (2.38), where η_i and γ_i are coefficients from the canonical form (2.37). Let us show that they are the quotient and the remainder of the division of the polynomial $\alpha(s)$ by $\beta(s)$, respectively. We carry out the Laplace transformation under zero initial conditions for system (2.37′). Then, preserving the same notation for the Laplace transformations of coordinates as for the preimages, we obtain relations

$$(sI - A_{11})x' = A_{12}y \Rightarrow x' = (sI - A_{11})^{-1}A_{12}y.$$

It follows from the last row of (2.37′) that

$$s^r y = -\overline{\eta}x' - \sum_{j=1}^{r} \gamma_j s^{j-1} y + u.$$

Hence

$$(s^r + \gamma_r s^{r-1} + \cdots + \gamma_1)y + \overline{\eta}x' = \varphi(s)y + \overline{\eta}(sI - A_{11})^{-1}A_{12}y = u.$$

Taking into account the explicit representation for A_{11} and A_{12}, we obtain

$$\overline{\eta}(sI - A_{11})^{-1}A_{12} = \frac{\psi(s)}{\beta(s)},$$

and then we have

$$\varphi(s)y + \frac{\psi(s)}{\beta(s)}y = u.$$

Since $y = \frac{\beta(s)}{\alpha(s)}u$, it follows that $u = \frac{\alpha(s)}{\beta(s)}y$. Consequently, $\varphi(s)$ is the quotient of the division of $\alpha(s)$ by $\beta(s)$ and $\psi(s)$ is the remainder. The theorem is proved. □

2.3.2 Zero dynamics of vector systems

We shall indicate now the algorithm of reduction of a vector system to a canonical form similar to form (2.37′). We shall assume that system (2.2) is in the general position, the dimensions of the input and output coincide (i.e., the system is square). In addition, suppose that the definition of the relative order of the vector system given by Isidori [60] is valid for system (2.2).

Definition 2.27 (by Isidori). The vector $r = (r_1, \dots, r_l)$ is a *relative order vector* of system (2.2) if the conditions

(1) $C_i A^j B = 0$, $j = 1, \dots, r_i - 2$; $C_i A^{r_i - 1} B \neq 0$ for all $i = 1, \dots, l$,

(2) $\det H(r_1, \dots, r_l) = \det \begin{pmatrix} C_1 A^{r_1-1} B \\ \cdots \\ C_l A^{r_l-1} B \end{pmatrix} \neq 0$,

where C_i are rows of the matrix C, are fulfilled.

Condition (1) of the definition means that the output derivatives $y_i = C_i x$ up to the order $(r_i - 1)$ inclusive do not depend explicitly on the input u and the r_ith derivative depends explicitly on u.

Condition (2) means that the matrix $H(r_1, \dots, r_l) = H(r)$ in the control $u(t) \in \mathbb{R}^l$ in the equations for the r_ith output derivatives is nondegenerate, i.e.,

$$\begin{pmatrix} y_1^{(r_1)} \\ y_2^{(r_2)} \\ \vdots \\ y_l^{(r_l)} \end{pmatrix} = \begin{pmatrix} C_1 A^{r_1} \\ C_2 A^{r_2} \\ \vdots \\ C_l A^{r_l} \end{pmatrix} x + H(r)u.$$

It is important to emphasize here that the conditions of the Isidori definition may be mutually incompatible for a square linear stationary system of the general position [17]. The following example illustrates the last statement.

Example 2.28. Consider a system of general position

$$\begin{cases} \dot{x}_1 = x_2 + u_1 \\ \dot{x}_2 = x_3 \\ \dot{x}_3 = u_2, \end{cases}$$

$$y_1 = x_1, \quad y_2 = x_1 + x_2.$$

In this case $l = 2$. We find r_1 and r_2 from condition (1) of the definition:

$$C_1 B = (1, 0) \Rightarrow r_1 = 1$$
$$C_2 B = (1, 0) \Rightarrow r_2 = 1.$$

However, for the vector $r = (1,1)$ the matrix $H(r)$ has the form

$$H(1,1) = \begin{pmatrix} 1 & 0 \\ 1 & 0 \end{pmatrix}, \quad \det H(1,1) = 0,$$

i.e., conditions (1) and (2) from Definition 2.27 are incompatible. Consequently, for the system in question the relative order in the sense of Isidori is not defined.

Let us show that in the case where the definition in the sense of Isidori is fulfilled, system (2.2) can be reduced to a special canonical form with the isolation of zero dynamics, namely, the system decomposes into two parts, the first of which describes the zero dynamics of the system, its input being the output of the system $y(t)$, and the second part consists of the derivatives of different orders of the outputs y_i. We divide the transformation into several stages. We shall begin with the proof of an auxiliary lemma.

Lemma 2.29. *Suppose that system* (2.2) *is observable,* $\operatorname{rank} C = l$, $\operatorname{rank} B = l$, *and the Isidori conditions are fulfilled. Then the rows* C_1, C_1A, $\dots$, $C_1A^{r_1-1}$, C_2, C_2A, $\dots$, $C_2A^{r_2-1}$, C_3, $\dots$, $C_lA^{r_l-1}$ *are linearly independent.*

Proof. Let us show that the set of vectors $C_1, C_1A, \dots, C_lA^{r_l-1}$ is linearly independent. Suppose the contrary, i.e., that these vectors are linearly dependent. Then there exist numbers $\gamma_1^1, \gamma_1^2, \dots, \gamma_1^{r_1}, \gamma_2^1, \dots, \gamma_l^{r_l}$, which are not all zero, such that

$$\sum_{i=1}^{l}\sum_{j=1}^{r_i} \gamma_i^j C_i A^{j-1} = 0. \tag{2.39}$$

We postmultiply this identity by the matrix B. Taking into account condition (1) of the definition of the relative order, we obtain

$$\gamma_1^{r_1} C_1 A^{r_1-1} B + \gamma_2^{r_2} C_2 A^{r_2-1} B + \dots + \gamma_l^{r_l} C_l A^{r_l-1} B = 0.$$

It follows from condition (2) of the definition that the rows $C_i A^{r_i-1} B$ are linearly independent, and therefore $\gamma_i^{r_i} = 0$ for all $i = 1, \dots, l$. In this way (2.39) assumes the form

$$\sum_{i=1}^{l}\sum_{j=1}^{r_i-1} \gamma_i^j C_i A^{j-1} = 0.$$

Now we postmultiply this identity by AB. Taking into account conditions (1) and (2), we find that $\gamma_i^{r_i-1} = 0$ for all $i = 1, \dots, l$. (In this case, if $r_j = 1$ for some j, then it may turn out that not all terms enter into the linear combination from $C_i A^{r_i-1} B$.) Continuing a similar reasoning, we find that all $\gamma_i^j = 0$. The lemma is proved. □

It follows from Lemma 2.29 that we can take the variables

$$\begin{array}{llll} y_1^1 = C_1 x, & y_1^2 = C_2 x, & \dots, & y_1^l = C_l x, \\ y_2^1 = C_1 A x, & y_2^2 = C_2 A x, & \dots, & y_2^l = C_l A x, \\ \vdots & \vdots & \vdots & \vdots \\ y_{r_1}^1 = C_1 A^{r_1 - 1} x, & y_{r_2}^2 = C_2 A^{r_2 - 1} x, & \dots, & y_{r_l}^l = C_l A^{r_l - 1} x \end{array} \tag{2.40}$$

as a part of new coordinates. We denote by $|r| = r_1 + \cdots + r_l$ the length of the vector r. Then (2.40) yields $|r|$ coordinates. The other $n - |r|$ coordinates we choose arbitrarily in the way that the transformation of coordinates becomes nondegenerate.

Let us introduce the following notation:

$$\begin{pmatrix} y_1^1 \\ \vdots \\ y_{r_1}^1 \end{pmatrix} = \bar{y}_1, \dots, \begin{pmatrix} y_1^l \\ \vdots \\ y_{r_l}^l \end{pmatrix} = \bar{y}_l$$

and denote the auxiliary coordinates as $\bar{x}' \in \mathbb{R}^{n-|r|}$. In the new coordinates the system assumes the form

$$\begin{aligned} &\dot{\bar{x}}' = \bar{A}_{11} \bar{x}' + \sum_{i=1}^{l} \bar{A}_{12}^i \bar{y}_i + B_1 u, \\ &\begin{cases} \dot{y}_1^1 = y_2^1 \\ \dot{y}_2^1 = y_3^1 \\ \vdots \\ \dot{y}_{r_1 - 1}^1 = y_{r_1}^1, \end{cases} \\ &\begin{cases} \dot{y}_1^2 = y_2^2 \\ \vdots \\ \dot{y}_{r_2 - 1}^2 = y_{r_2}^2, \end{cases} \\ &\cdots\cdots\cdots\cdots \\ &\begin{cases} \dot{y}_1^l = y_2^l \\ \vdots \\ \dot{y}_{r_l - 1}^l = y_{r_l}^l, \end{cases} \\ &\left\{ \begin{pmatrix} \dot{y}_{r_1}^1 \\ \dot{y}_{r_2}^2 \\ \vdots \\ \dot{y}_{r_l}^l \end{pmatrix} = \bar{A}_{21} \bar{x}' + \sum_{i=1}^{l} \bar{A}_{22}^i \bar{y}_i + H(r_1, \dots, r_l) u, \right. \end{aligned} \tag{2.41}$$

where $\bar{A}_{11}, \bar{A}_{21}, \bar{A}^i_{12}, \bar{A}^i_{22}$ are constant matrices of the corresponding dimensions which are uniquely defined by the transformation of coordinates and by the parameters of the system.

Note that the matrix $H(r) = H(r_1, \dots, r_l)$ is nondegenerate, and therefore we can use the nondegenerate transformation of the first $(n - |r|)$ coordinates

$$\tilde{x}' = \bar{x}' - B_1 H^{-1}(r) \begin{pmatrix} y^1_{r_1} \\ \vdots \\ y^l_{r_l} \end{pmatrix}$$

to achieve, at the 2nd step, the situation where the first $(n - |r|)$ coordinates will not explicitly depend on $u(t)$. In the new coordinates the first $(n - |r|)$ equations of the system assume the form

$$\dot{\tilde{x}}' = \widetilde{A}_{11}\tilde{x}' + \sum_{i=1}^{l} \widetilde{A}^i_{12}\bar{y}_i, \tag{2.42}$$

where the matrices $\widetilde{A}_{11}$ and $\widetilde{A}^i_{12}$ are uniquely defined by the indicated change of coordinates.

In representation (2.42) the coordinates $\tilde{x}'$ explicitly depend not only on the outputs y_i but also on their derivatives (i.e., on full vectors $\bar{y}_i$). Let us show that we can get rid of these derivatives.

For this purpose we shall show that in representation (2.42) we can remove $y^1_{r_1} = y_1^{(r_1-1)}$ which is the leading $(r_i - 1)$th derivative of the first output. Let us write (2.42) in greater detail:

$$\dot{\tilde{x}}' = \widetilde{A}_{11}\tilde{x}' + \sum_{i=2}^{l} \widetilde{A}^i_{12}\bar{y}_i + \sum_{j=1}^{r_1} \left(\widetilde{A}^1_{12}\right)_j y^1_j, \tag{2.43}$$

where $\left(\bar{A}^1_{12}\right)_j$ are columns of the matrix $\widetilde{A}^1_{12}$. Taking into account the fact that $\dot{y}^1_{r_1 \; 1} = y^1_{r_1}$, we make a change

$$\hat{x}' = \tilde{x}' - \left(\widetilde{A}^1_{12}\right)_{r_1} y^1_{r_1-1} \tag{2.44}$$

and then

$$\begin{aligned} \dot{\hat{x}}' &= \dot{\tilde{x}}' - \left(\widetilde{A}^1_{12}\right)_{r_1} \dot{y}^1_{r_1-1} \\ &= \widetilde{A}_{11}\hat{x}' + \widetilde{A}_{11} \cdot \left(\widetilde{A}^1_{12}\right)_{r_1} y^1_{r_1-1} + \sum_{i=2}^{l} \widetilde{A}^i_{12}\bar{y}_i + \sum_{j=1}^{r_1} \left(\widetilde{A}^1_{12}\right)_j y^1_j - (\tilde{A}^1_{12})_{r_1} \cdot y^1_{r_1} \\ &= \widetilde{A}_{11}\hat{x}' + \sum_{i=2}^{l} \left(\widetilde{A}^i_{12}\right) \bar{y}_i + \sum_{j=1}^{r_1-1} \left(\hat{A}^1_{12}\right) y^1_j. \end{aligned} \tag{2.45}$$

Equation (2.45) differs from (2.43) by the fact that $y^1_{r_1}$ does not appear in it, with the columns changing for y^1_j $(j = 1, \dots, r_1 - 1)$. Thus the change of coordinates allows us to get rid of $y^1_{r_1}$ in the first $(n - |r|)$-dimensional part of the system.

At the next step, we can get rid, by analogy, of $y^1_{r_1-1}$ (taking into account that $\dot{y}^1_{r_1-2} = y^1_{r_1-1}$), then of $y^1_{r_1-2}$, and so on, until the vector $\bar{y}_1$ will contain only the coordinate $y^1_1 = y_1$, i.e., the first output of the system.

Then, using the same scheme, we can remove successively the derivatives of the other outputs of the system. At the indicated changes of coordinates in the remaining part of the system the equations for y^i_j $(i = 1, \dots, l,\ j = 1, \dots, r_i - 1)$ will not change, only the equations for the leading derivatives y_i will change, i.e., for coordinates $y^i_{r_i}$.

We have thus proved the following theorem.

Theorem 2.30. *Suppose that the linear stationary square system* (2.2) *is in the general position and the vector* $r = (r_1, \dots, r_l)$ *of the relative order in the sense of Isidori is defined. Then, by means of nondegenerate transformation the system can be reduced to the form*

$$
\begin{aligned}
\dot{x}' &= A_{11}x' + A_{12}y \\
\dot{y}^1_i &= y^1_{i+1}, \quad i = 1, \dots, r_1 - 1 \\
\dot{y}^2_i &= y^2_{i+1}, \quad i = 1, \dots, r_2 - 1 \\
&\vdots \\
\dot{y}^l_i &= y^l_{i+1}, \quad i = 1, \dots, r_l - 1 \\
\begin{pmatrix} \dot{y}^1_{r_1} \\ \vdots \\ \dot{y}^l_{r_l} \end{pmatrix} &= A_{21}x' + \sum_{i=1}^{l} A^i_{22}\bar{y}_i + H(r)u
\end{aligned}
\tag{2.46}
$$

which is the canonical representation of the system with the isolation of zero dynamics.

Corollary 2.31. *The zero dynamics of the system corresponds to the* $(n - |r|)$*-dimensional first part of the system and is described by the equation*

$$\dot{x}' = A_{11}x'.$$

Corollary 2.32. *It is easy to verify that the definition of the vector relative order in the sense of Isidori for the vector* $r = (r_1, \dots, r_l)$ *holds for* (2.46). *Consequently, the reduction of the system of general order* (2.2) *to form* (2.46) *is a necessary and sufficient condition for the fulfillment of the conditions of the definition of relative order.*

2.4 Nonstationary linear systems

We shall describe now the canonical representation for a linear nonstationary system [87].

In what follows we shall consider a homogeneous system of the form

$$\begin{cases} \dot{x}(t) = A(t)x(t) \\ \quad y = C(t)x(t). \end{cases} \tag{2.47}$$

Let us consider a transformation of coordinates with a matrix $L(t)$

$$x(t) = L(t)\bar{x}(t). \tag{2.48}$$

Let $L(t) \in \mathbb{R}^{n\times n}$ be invertible for all t, $|\det L(t)| \geq \text{const} > 0$, $L(t)$, $\dot{L}(t)$, and $L^{-1}(t)$ are continuous and bounded matrices (which realize the *Lyapunov transformation*).

The following statement holds true [87].

Lemma 2.33. *Suppose that system* (2.47) *is uniformly differentially observable* (*in the sense of Definition 2.10. Then there exists a nondegenerate* (*Lyapunov's*) *transformation* (2.48) *which reduces system* (2.47) *to the form*

$$\begin{cases} \dot{\bar{x}}(t) = \bar{A}(t)\bar{x}(t) \\ y(t) = \bar{C}(t)\bar{x}(t), \end{cases} \tag{2.49}$$

where

$$\bar{A}(t) = L^{-1}(t)A(t)L(t) - L^{-1}(t)\dot{L}(t) = \begin{pmatrix} \bar{A}_{00}(t) & \bar{A}_{01} & \dots & \bar{A}_{0l} \\ \bar{A}_{10}(t) & \bar{A}_{11} & \dots & \bar{A}_{1l} \\ \dots & \dots & \dots & \dots \\ \bar{A}_{l0}(t) & \bar{A}_{l1} & \dots & \bar{A}_{ll} \end{pmatrix}, \tag{2.50}$$

$\bar{C}(t) = C(t)L(t) = (\bar{C}_1(t); 0)$, $\bar{C}_1(t) \in \mathbb{R}^{l\times l}$, $\bar{C}_1(t) > 0$ (*is positive definite*).

The matrices $\bar{A}_{ij}$ for $j \geq 1$ from the block representation of the matrix $\bar{A}(t)$ do not depend on t and have the form

$$\bar{A}_{ii} = \begin{pmatrix} 0 & 1 & 0 & \dots & 0 \\ 0 & 0 & 1 & \dots & 0 \\ \dots & \dots & \dots & \dots & \dots \\ 0 & 0 & 0 & \dots & 1 \\ 0 & 0 & 0 & \dots & 0 \end{pmatrix}, \quad \bar{A}_{ii} \in \mathbb{R}^{(\nu_i-1)\times(\nu_i-1)}, \quad i = 1,\dots,l,$$

$$\bar{A}_{0i} = [e_i; 0], \quad e_i \in \mathbb{R}^{l\times 1} \text{ is the } i\text{th basis vector}, \quad \bar{A}_{0i} \in \mathbb{R}^{l\times(\nu_i-1)},$$

$$\bar{A}_{ij} = 0, \quad \bar{A}_{ij} \in \mathbb{R}^{(\nu_i-1)\times(\nu_j-1)}, \quad i \neq j, \ i,j = 1,\dots,l.$$

Here, as before, ν_i are observability indices, $\nu_1 + \nu_2 + \cdots + \nu_l = n$. The matrices $\bar{A}_{i0}(t) \in \mathbb{R}^{(\nu_i - 1)\times l}$, $i = 1, \ldots, l$, and $\bar{A}_{00}(t) \in \mathbb{R}^{l\times l}$ include all components of the matrix A dependent on time.

Remark 2.34. If the indicated transformation is realized for system (2.1), then the matrix $B(t)$ passes into $\bar{B}(t) = L^{-1}(t)B(t)$.

Remark 2.35. In the new basis system (2.47) can be written as a collection $(l + 1)$ of subsystems, i.e., in the form

$$\begin{cases} \begin{cases} \dot{x}_0 = \bar{A}_{00}(t)x_0 + \sum_{i=1}^{l} \bar{A}_{0i} x_i \\ \dot{x}_i = \bar{A}_{ii} x_i + \bar{A}_{i0}(t)x_0, \quad i = 1, \ldots, l \end{cases} \\ y = \bar{C}_1(t)x_0. \end{cases}$$

The indicated canonical representation is similar to the observable Luenberger representation and generalizes the observable Tuel representation for linear time-dependent systems [106, 117].

The procedure of construction of the Lyapunov transformation of $L(t)$ for reducing the system to the described canonical form can be realized according to the algorithm described in [87].

(a) For any time moment $t \geq 0$ the observability matrix of full rank (2.6)

$$N(t) = \begin{pmatrix} Q_1(t) \\ \vdots \\ Q_n(t) \end{pmatrix}, \quad Q_1(t) = C(t), \quad Q_{i+1}(t) = Q_i(t)A(t) + \dot{Q}_i(t)$$

can be uniquely defined. From the rows of this matrix we choose basis rows in accordance with the procedure described in detail for stationary systems (Technique 2). Moreover, we assume that the set of rows satisfying this procedure for $t = 0$ satisfies it for all $t \geq 0$ as well (and is a basis one for $t \geq 0$).

We denote by $q_{ij}(t)$ the jth row of the matrix $Q_i(t)$. The observability indices ν_k, $k = 1, \ldots, l$, mean the maximal value for the first index of the row $q_{ij}(t)$, i.e., $1 \leq i \leq \nu_j$. In this case, the index j corresponds to the jth output of the system. In particular, $q_{1j}(t) = C_j(t)$ are rows of the initial matrix $C(t)$.

Let us compose a nondegenerate matrix $\hat{M}(t) \in \mathbb{R}^{n\times n}$ from the chosen basis rows ordering them as follows:

$$\hat{M}(t) = \begin{pmatrix} \hat{M}_1(t) \\ \hline \hat{M}_2(t) \end{pmatrix},$$

where

$$\hat{M}_1(t) = \begin{pmatrix} q_{\nu_1,1} \\ q_{\nu_2,2} \\ \vdots \\ q_{\nu_l,l} \end{pmatrix},$$

and $\hat{M}_2(t)$ contains the remaining rows $q_{ij}(t)$, $j = 1, \dots, l$; $i = 1, \dots, \nu_j - 1$.

(b) Let $h_i(t)$, $i = 1, \dots, l$ be the ith columns of the matrix $\hat{M}^{-1}(t)$.

(c) Let us define the vectors $l_{ij} \in \mathbb{R}^{n\times 1}$, $i = 1, \dots, l$, $j = 1, \dots, \nu_i$:

$$l_{i1}(t) = h_i(t), \quad i = 1, \dots, l$$
$$l_{ij}(t) = A(t)l_{i,j-1}(t) - \dot{l}_{i,j-1}(t), \quad i = 1, \dots, l; \quad j = 2, \dots, \nu_i.$$

(d) Let us construct now matrices

$$L_0(t) = (l_{1,\nu_1}; l_{2,\nu_2}, \dots, l_{l,\nu_l})$$
$$L_i(t) = (l_{i,\nu_1-1}; l_{i,\nu_i-2}, \dots, l_{i,1}), \quad i = 1, \dots, l.$$

(e) Finally, the matrix $L(t) \in \mathbb{R}^{n\times n}$ which realizes the required Lyapunov transformation assumes the form

$$L(t) = (L_0(t), L_1(t), \dots, L_l(t)).$$

Conclusion

In Chapter 2 we consider the concepts of observability and identifiability for linear dynamical systems (Definitions 2.2–2.3. The observability criteria of Kalman (Theorem 2.5) and Rosenbrock (Theorem 2.11) are given for linear stationary systems.

The following canonical forms are given for linear stationary systems: the Kalman decomposition, observability canonical forms (for scalar and vector outputs), the Luenberger canonical form, the canonical form with isolation of zero dynamics. The Luenberger canonical form is given for linear nonstationary systems.

Chapter 3

Observers of full-phase vector for fully determined linear systems

In this chapter we consider fully determined linear stationary systems

$$\begin{cases} \dot{x} = Ax + Bu \\ y = Cx, \end{cases} \tag{3.1}$$

where $x \in \mathbb{R}^n$ is an unknown phase vector, $u(t) \in \mathbb{R}^m$, $y(t) \in \mathbb{R}^l$ are known input and output of the system, A, B, and C are known constant matrices of the corresponding dimensions, C and B being matrices of full rank.

3.1 Full-dimensional observers

Let us consider a problem on the construction of an asymptotic observer which forms an estimate of the unknown phase vector $x(t)$. We shall begin with the description of the most simple methods of constructing observers whose dimension coincides with that of system (3.1) (the so-called *full-dimensional observers*).

Let us prove the following fundamental statement which is very important in the observation theory. The following theorem is valid.

Theorem 3.1. *Let the pair $\{C, A\}$ be observable. Then, for any polynomial $\varphi_n(s) = \gamma_1 + \gamma_2 s + \cdots + \gamma_n s^{n-1} + s^n$ with real coefficients γ_i, $s \in \mathbb{C}$, there exists a matrix $L \in \mathbb{R}^{n \times l}$ such that $\varphi_n(s)$ is a characteristic polynomial of the matrix $A_L = A - LC$, i.e,*

$$\det(sI - A_L) = \varphi_n(s).$$

Remark 3.2. Actually, Theorem 3.1 states that for the observable pair $\{C, A\}$, by the choice of the matrix L the spectrum of the matrix A_L may be defined arbitrarily (with the natural restriction that the spectrum must be symmetric relative to the real axis in $\mathbb{C}$).

Proof. We can prove the statement in any specially chosen, say, canonical basis. Indeed, let the transformation of coordinates with the matrix P, i.e., $\bar{x} = Px$, transfer the pair $\{C, A\}$ into a pair $\{\bar{C}, \bar{A}\}$, where $\bar{C} = CP^{-1}$, $\bar{A} = PAP^{-1}$. Then the characteristic polynomials of the matrices $A_L = A - LC$ and $\bar{A}_L = \bar{A} - \bar{L}\bar{C}$, where

$\bar{L} = PL$, coincide, and, consequently, by choosing the matrix $\bar{L}$ we find the matrix $L = P^{-1}\bar{L}$.

We begin with considering the case $l - 1$. If the pair $\{C, A\}$ is observable, then, with the use of a nondegenerate transformation, we can transfer it to the second observable canonical representation (2.19)

$$A = \begin{pmatrix} 0 & 0 & \dots & 0 & -\alpha_1 \\ 1 & 0 & \dots & 0 & -\alpha_2 \\ \dots & \dots & \dots & \dots & \dots \\ 0 & 0 & \dots & 1 & -\alpha_n \end{pmatrix}; \quad C = (0, \dots, 0, 1).$$

Let $L = (l_1, \dots, l_n)^\top$. Then

$$A_L = A - LC = \begin{pmatrix} 0 & 0 & \dots & 0 & -(\alpha_1 + l_1) \\ 1 & 0 & \dots & 0 & -(\alpha_2 + l_2) \\ 0 & 1 & \dots & 0 & -(\alpha_3 + l_3) \\ \dots & \dots & \dots & \dots & \dots \\ 0 & 0 & \dots & 1 & -(\alpha_n + l_n) \end{pmatrix}$$

and $\det(sI - A_L) = s^n + (\alpha_n + l_n)s^{n-1} + \dots + (\alpha_1 + l_1)$. Choosing $l_i = \gamma_i - \alpha_i$, we obtain

$$\det(sI - A_L) = \varphi_n(s).$$

In this way we have proved the statement for $l = 1$.

Let us now consider the general case $l > 1$. We begin with proving an auxiliary statement.

Lemma 3.3. *Suppose that the pair $\{C, A\}$ is observable and C_i are rows of the matrix C. Then, for any row $C_i \neq 0$, $i = 1, \dots, l$, there exists a matrix $L_i \in \mathbb{R}^{n \times l}$ such that the pair $\{C_i, A - L_i C\}$ is also observable.*

Proof. Without loss of generality, we consider $i = 1$. Since the pair $\{C, A\}$ is observable, $\operatorname{rank} N(C, A) = n$ and therefore among the rows of the matrix $N(C, A)$ there are n linearly independent rows. We choose them using Technique 1 (see Chapter 2). They are the vectors

$$\begin{gathered} \{C_1, \dots, C_1 A^{\nu_1 - 1}, C_2, \dots, C_2 A^{\nu_2 - 1}, \dots, C_k, \dots, C_k A^{\nu_k - 1}\}, \\ \nu_1 + \dots + \nu_k = n, \ k \le l. \end{gathered} \tag{3.2}$$

Let us determine the matrix $P \in \mathbb{R}^{n \times n}$ whose rows are these vectors in the indicated order. Let us determine the matrix $S \in \mathbb{R}^{n\times l}$

$$S = \begin{pmatrix} 0_{\nu_1-1} \\ -e_2 \\ 0_{\nu_2-1} \\ -e_3 \\ \vdots \\ 0_{\nu_{k-1}-1} \\ -e_{k-1} \\ 0_{\nu_k} \end{pmatrix} \begin{matrix} \left.\vphantom{\begin{matrix}a\\b\end{matrix}}\right\} \nu_1 \\ \left.\vphantom{\begin{matrix}a\\b\end{matrix}}\right\} \nu_2 \\ \vdots \\ \left.\vphantom{\begin{matrix}a\\b\end{matrix}}\right\} \nu_{k-1} \\ \} \nu_k \end{matrix}, \quad 0_p \in \mathbb{R}^{p\times l},$$

e_i being the ith row of the identity $l \times l$ matrix. We shall show that the matrix $L_1 = P^{-1}S$ satisfies the conditions of Lemma 3.3.

We write the matrix $S = PL_1$ in the form

$$S = \begin{pmatrix} C_1 \\ C_1 A \\ \vdots \\ C_k A^{\nu_k-1} \end{pmatrix} \cdot L_1 = \begin{pmatrix} 0 \\ \vdots \\ 0 \\ -e_2 \\ 0 \\ \vdots \\ 0 \\ -e_3 \\ \vdots \\ -e_{k-1} \\ 0 \\ \vdots \\ 0 \end{pmatrix}.$$

This matrix relation corresponds to the following set of vector relations

$$\begin{cases} C_1L_1 = 0, & C_1AL_1 = 0, \ \dots, & C_1A^{\nu_1-1}L_1 = -e_2 \\ C_2L_1 = 0, & C_2AL_1 = 0, \ \dots, & C_2A^{\nu_2-1}L_1 = -e_3 \\ \vdots & & \\ C_{k-1}L_1 = 0, & C_{k-1}AL_1 = 0, \ \dots, & C_{k-1}A^{\nu_{k-1}-1}L_1 = -e_k \\ C_kL_1 = 0, & C_kAL_1 = 0, \ \dots, & C_kA^{\nu_k-1}L_1 = 0. \end{cases} \tag{3.3}$$

In order to prove that the pair $\{C_1, A - L_1C\}$ is observable, we have to prove that the rows

$$C_1, \ C_1(A - L_1C), \ C_1(A - L_1C)^2, \ \dots, \ C_1(A - L_1C)^{n-1}$$

are linearly independent. Using relations (3.3), we find that

$$\begin{aligned}
C_1 &= C_1\\
C_1(A-L_1C) &= C_1A\\
C_1(A-L_1C)^2 &= C_1A^2\\
&\vdots\\
C_1(A-L_1C)^{\nu_1-1} &= C_1A^{\nu_1-1}\\
C_1(A-L_1C)^{\nu_1} &= CA^{\nu_1-1}(A-L_1C) = CA^{\nu_1}+e_2C = CA^{\nu_1}+C_2 \qquad (3.4)\\
C_1(A-L_1C)^{\nu_1+1} &= CA^{\nu_1}+C_2(A-L_1C) = C_2A+[\dots]\\
C_1(A-L_1C)^{\nu_1+2} &= C_2A^2+[\dots]\\
&\vdots\\
C_1(A-L_1C)^{\nu_1-1} &= C_kA^{\nu_k-1}+[\dots].
\end{aligned}$$

In this system the brackets $[\dots]$ are used to denote the linear combinations of the preceding vectors.

Since (3.2) is a set of linearly independent vectors, vectors (3.4) are also linearly independent. The lemma is proved. □

Let us now pass to the proof of the theorem itself. Let the pair $\{C, A\}$ be observable. Then we consider a matrix L_1 such that the pair $\{C_1, A - L_1C\}$ is observable. Since for the case $l = 1$ the statement of the theorem has been proved, there exists a matrix $\bar{L}' \in \mathbb{R}^{n\times 1}$ such that the matrix

$$A_L = (A - L_1C) - \bar{L}'C_1$$

has a defined spectrum. Let us consider a matrix

$$L' = (\bar{L}'0\dots0) \in \mathbb{R}^{n\times l}.$$

Then the matrix A_L from condition (3.4) can be written as

$$A_L = (A - (L_1 + L')C) = A - LC,$$

where $L = L_1 + L'$. The theorem is proved. □

Remark 3.4. [to Lemma 3.3] The observability of the pair $\{C, A\}$ does not, generally speaking, implies the observability of the pair $\{C_i, A\}$ for any $i = 1, \dots, l$. Moreover, the matrix A may not be observable with respect to any vector $C' \in \mathbb{R}^{1\times n}$!

Let us consider an example.

Example 3.5. Suppose that we are given a system with matrices

$$A = \begin{pmatrix} 0 & 1 & 0 \\ 0 & 0 & 0 \\ 0 & 0 & 0 \end{pmatrix}, \quad C = \begin{pmatrix} 1 & 0 & 0 \\ 0 & 0 & 1 \end{pmatrix}.$$

Then the observability matrix for the pair $\{C, A\}$ has the form

$$N(C, A) = \begin{pmatrix} 1 & 0 & 0 \\ 0 & 0 & 1 \\ 0 & 1 & 0 \\ 0 & 0 & 0 \end{pmatrix}, \quad \operatorname{rank} N(C, A) = 3,$$

i.e., the pair $\{C, A\}$ is observable.

However, for any vector $C' = (c_1, c_2, c_3)$ the observability matrix

$$N(C', A) = \begin{pmatrix} c_1 & c_2 & c_3 \\ 0 & c_2 & 0 \\ 0 & 0 & 0 \end{pmatrix}$$

has a rank not higher than the second. Thus, the pair $\{C', A\}$ is not observable with respect to any vector C' including the vectors C_1 and C_2 which are the rows of the matrix C.

Let us consider now the problem of constructing an asymptotic full-dimensional observer for the linear stationary system (3.1). We can easily do it proceeding from Theorem 3.1.

Suppose that we are given a linear stationary system (3.1)

$$\begin{cases} \dot{x} = Ax + Bu \\ y = Cx. \end{cases}$$

We construct a *full-dimensional observer* for this system as

$$\dot{\tilde{x}} = A\tilde{x} + Bu - L(C\tilde{x} - y), \quad \tilde{x} \in \mathbb{R}^n, \tag{3.5}$$

where we choose the feedback matrix L from the condition that $A_L = A - LC$ is a Hurwitz matrix. This can always be done when the pair $\{C, A\}$ is observable. The observation error $e = x - \tilde{x}$ in this case satisfies the differential equation

$$\dot{e} = \dot{x} - \dot{\tilde{x}} = Ax + Bu - A\tilde{x} - Bu - L(C\tilde{x} - Cx) = A_L e,$$

and $e(t) \to 0$ as $t \to 0$ since A_L is a Hurwitz matrix. Moreover, since the estimation error e satisfies the linear stationary equation indicated above, it follows that $e(t) \to 0$ exponentially, the rate of convergence can be chosen arbitrarily with the use of the matrix L.

3.1.1 Algorithms of the synthesis of observers with the use of different canonical forms

We shall now indicate some explicit algorithms for the synthesis of feedback matrix L.

In the case where $l = 1$, it is easy to find the coefficients of the matrix L if system (3.1) is given in the second observable canonical form. The method of synthesis of L for this case is given in the proof of Theorem 3.1.

Let us consider the case where $l = 1$ and the system is given in the first observable canonical form (2.18)

$$A = \begin{pmatrix} 0 & 1 & 0 & \dots & 0 \\ 0 & 0 & 1 & \dots & 0 \\ \multicolumn{5}{c}{\dotfill} \\ 0 & 0 & 0 & \dots & 1 \\ -\alpha_1 & -\alpha_2 & -\alpha_3 & \dots & -\alpha_n \end{pmatrix}, \quad C = (1, 0, \dots, 0).$$

Since the indicated pair is observable, the existence of the vector L follows from what was proved above. However, we can also indicate an explicit method for finding L.

Let $L = (\tilde{l}_n, \dots, \tilde{l}_1)^\top$. Then

$$A_L = \begin{pmatrix} -\tilde{l}_n & 1 & 0 & \dots & 0 & 0 \\ -\tilde{l}_{n-1} & 0 & 1 & \dots & 0 & 0 \\ \multicolumn{6}{c}{\dotfill} \\ -\tilde{l}_3 & 0 & 0 & \dots & 1 & 0 \\ -\tilde{l}_2 & 0 & 0 & \dots & 0 & 1 \\ -(\tilde{l}_1 + \alpha_1) & -\alpha_2 & -\alpha_3 & \dots & -\alpha_{n-1} & -\alpha_n \end{pmatrix}.$$

Let us find the characteristic polynomial of the matrix A_L

$$\det(sI - A_L) = \det \begin{pmatrix} (s + \tilde{l}_n) & -1 & 0 & \dots & 0 & 0 \\ \tilde{l}_{n-1} & s & -1 & \dots & 0 & 0 \\ \multicolumn{6}{c}{\dotfill} \\ \tilde{l}_3 & 0 & s & \dots & -1 & 0 \\ \tilde{l}_2 & 0 & 0 & \dots & s & -1 \\ (\tilde{l}_1 + \alpha_1) & \alpha_2 & \alpha_3 & \dots & \alpha_{n-1} & (s + \alpha_n) \end{pmatrix}$$

$$= (\tilde{l}_1 + \alpha_1) + \tilde{l}_2(\alpha_n + s) + \tilde{l}_3 \det \begin{pmatrix} s & -1 \\ \alpha_{n-1} & (\alpha_n + s) \end{pmatrix}$$

$$+ \tilde{l}_4 \det \begin{pmatrix} s & -1 & 0 \\ 0 & s & -1 \\ \alpha_{n-2} & \alpha_{n-1} & (\alpha_n + s) \end{pmatrix} + \dots + \tilde{l}_{n-1} \det \begin{pmatrix} s & -1 & \dots & 0 \\ 0 & s & \dots & 0 \\ \multicolumn{4}{c}{\dotfill} \\ \alpha_3 & \alpha_4 & \dots & (\alpha_n + s) \end{pmatrix}$$

$$+(\tilde{l}_n + s)\det\begin{pmatrix} s & -1 & \dots & 0 \\ 0 & s & \dots & 0 \\ \dots & \dots & \dots & \dots \\ \alpha_2 & \alpha_3 & \dots & (\alpha_n + s) \end{pmatrix}$$

$$= (\tilde{l}_1 + \alpha_1) + \tilde{l}_2 p_2(s) + \tilde{l}_3 p_3(s) + \dots + \tilde{l}_{n-1} p_{n-1}(s) + (\tilde{l}_n + s) p_n(s).$$

Every polynomial $p_i(s) = s^{i-1} + \dots$, and therefore the representation

$$\psi_i = \tilde{l}_i + \sum_{j=i+1}^{n} \varphi_{ij}\tilde{l}_j$$

holds for the coefficients of the characteristic polynomial $\det(sI - A_L) = s^n + \psi_n s^{n-1} + \psi_{n-1}s^{n-2} + \dots + \psi_1$.

Setting ψ_i equal to the given coefficients γ_i, we obtain a linear system relative to the unknown coefficients $\tilde{l}_i$. In this case, the matrix of this system will be triangular with identities on the leading diagonal

$$\begin{pmatrix} 1 & \varphi_{1,2} & \dots & \varphi_{1,n-2} & \varphi_{1,n-1} & \varphi_{1,n} \\ \dots & \dots & \dots & \dots & \dots & \dots \\ 0 & 0 & \dots & 1 & \varphi_{n-2,n-1} & \varphi_{n-2,n} \\ 0 & 0 & \dots & 0 & 1 & \varphi_{n-1,n} \\ 0 & 0 & \dots & 0 & 0 & 1 \end{pmatrix} \begin{pmatrix} \tilde{l}_1 \\ \vdots \\ \tilde{l}_n \end{pmatrix} = \begin{pmatrix} \gamma_1 \\ \vdots \\ \gamma_n \end{pmatrix}.$$

In this way we can find the required feedback vector for the case where the pair $\{C, A\}$ is given in the first canonical representation.

3.1.2 Synthesis of observers with the use of the first canonical representation of vector systems

Proceeding from the scalar case, we can propose a method of synthesis of the matrix L for the vector case $l > 1$. Suppose that the system is given in the first canonical representation for vector systems (2.22), (2.23), i.e.,

$$A = \begin{pmatrix} A_{11} & 0 & \dots & 0 \\ A_{21} & A_{22} & \dots & 0 \\ \vdots & \vdots & \ddots & \vdots \\ A_{k1} & A_{k2} & \dots & A_{kk} \end{pmatrix}, \quad A_{ii} = \begin{pmatrix} 0 & 1 & 0 & \dots & 0 \\ 0 & 0 & 1 & \dots & 0 \\ \dots & \dots & \dots & \dots & \dots \\ 0 & 0 & 0 & \dots & 1 \\ * & * & * & \dots & * \end{pmatrix} \in \mathbb{R}^{\nu_i \times \nu_i},$$

$$C = \begin{pmatrix} \bar{C}_1 & 0 & \dots & 0 \\ 0 & \bar{C}_2 & \dots & 0 \\ \vdots & \vdots & \ddots & \vdots \\ 0 & 0 & \dots & \bar{C}_k \\ \bar{C}_{k+1}^1 & \bar{C}_{k+1}^2 & \dots & \bar{C}_{k+1}^k \end{pmatrix},$$

$$\bar{C}_i = (1, 0, \dots, 0) \in \mathbb{R}^{1\times\nu_i}, \quad \bar{C}^i_{k+1} \in \mathbb{R}^{(l-k)\times\nu_i}, \quad \nu_1 + \dots + \nu_k = n.$$

Since every pair $\{\bar{C}_i, A_{ii}\}$ is given in the first canonical form for the scalar system, it follows that, using the algorithm described above, we can find for each pair a matrix $\bar{L}_i \in \mathbb{R}^{\nu_i\times 1}$ such that the matrix $A^L_{ii} = A_{ii} - \bar{L}_i\bar{C}_i$ has a given spectrum. The general matrix $L \in \mathbb{R}^{n\times l}$ can be chosen in the form

$$L = \begin{pmatrix} \bar{L}_1 & 0 & \dots & 0 & 0 \\ 0 & \bar{L}_2 & \dots & 0 & 0 \\ \dots & \dots & \dots & \dots & \dots \\ 0 & 0 & \dots & \bar{L}_k & 0 \end{pmatrix}.$$

In this case

$$A_L = A - LC = \begin{pmatrix} A^L_{11} & 0 & \dots & 0 \\ A_{21} & A^L_{22} & \dots & 0 \\ \vdots & \vdots & \ddots & \vdots \\ A_{k1} & A_{k2} & \dots & A^L_{kk} \end{pmatrix},$$

and, consequently, the characteristic polynomial $\alpha(s)$ of the matrix A_L is the product of the characteristic polynomials of the matrices A^L_{ii} and can be made a Hurwitz polynomial (with any preassigned index of stability).

Note that the indicated method of synthesis imposes constraints on the choice of the spectrum of A_L because the real characteristic polynomials of A^L_{ii} may not admit of some combinations of complex-conjugate roots for the spectrum of A_L. For instance, if $n = 2$ and the matrix A consists of two blocks (i.e., $k = 2$), then the characteristic polynomial

$$\alpha(s) = \det(sI - A_L) = \alpha_1(s)\alpha_2(s),$$

where $\alpha_i(s) = \det(sI - A^L_{ii})$ are first-order polynomials, and, consequently, for the indicated technique of synthesis of the matrix L the spectrum of the matrix A_L will necessarily be real.

In order to be able to obtain an arbitrary spectrum of the matrix A_L, it is necessary to construct the matrix L in a more general form.

3.1.3 Synthesis with the use of the Luenberger form

One more technique of synthesizing an observer is connected with the canonical Luenberger form for systems with vector output (2.33) when the system decomposes into l subsystems

$$\dot{x}_i = A_{ii}x_i + \sum_{j=1, j\neq i}^{l} \bar{a}_{ij}y_j + B_i u, \quad i = 1, \dots, l; \quad x_i \in \mathbb{R}^{\nu_i},$$

$$y_i = \bar{C}_i x_i, \quad \bar{C}_i = (0, \ldots, 0, 1), \quad A_{ii} = \begin{pmatrix} 0 & 0 & \ldots & 0 & * \\ 1 & 0 & \ldots & 0 & * \\ \ldots & \ldots & \ldots & \ldots & \ldots \\ 0 & 0 & \ldots & 1 & * \end{pmatrix}.$$

Since each pair $\{\bar{C}_i, A_{ii}\}$ is in the second canonical representation, in accordance with the algorithm given in the proof of Theorem 3.1 we find a vector $\bar{L}_i \in \mathbb{R}^{\nu_i \times 1}$ such that $A_{ii}^L = A_{ii} - \bar{L}_i \bar{C}_i$ is a Hurwitz matrix (and has a defined degree of stability).

We construct an observer in the form

$$\dot{\tilde{x}}_i = A_{ii}\tilde{x}_i + \sum_{j=1, j\neq i}^{l} \bar{a}_{ij} y_j + B_i u - \bar{L}_i(\bar{C}_i \tilde{x}_i - y_i), \quad i = 1, \ldots, l. \tag{3.6}$$

The observation errors $e_i = x_i - \tilde{x}_i$ for each subsystem satisfy the equations

$$\dot{e}_i = A_{ii}^L e_i,$$

and, consequently, $e_i(t) \to 0$ exponentially as $t \to \infty$.

Remark 3.6. As in the preceding case, we impose small constraints on the spectrum of the matrix of the system in the deviations in the large, the constraints being connected with the impossibility of defining certain spectra containing complex-conjugate values.

3.1.4 Synthesis of observers with the reconstructible pair $\{C, A\}$

Let us now consider the case where the pair $\{C, A\}$ is not fully observable. The following statement is valid for the pair $\{C, A\}$ being reconstructed.

Theorem 3.7. *Suppose that the pair $\{C, A\}$ is reconstructible. Then there exists a matrix $L \in \mathbb{R}^{n\times l}$ such that $A_L = A - LC$ is a Hurwitz matrix.*

Proof. Note, first of all, that it suffices to carry out the proof for a system written in an arbitrary canonical basis. Indeed, if, by means of the transformation of coordinates the pair $\{C, A\}$ is transferred into a pair $\{\bar{C}, \bar{A}\}$, where $\bar{C} = CP^{-1}$, $\bar{A} = PAP^{-1}$, and for the pair $\{\bar{C}, \bar{A}\}$ there exists a matrix $\bar{L}$ such that $\bar{A}_L = \bar{A} - \bar{L}\bar{C}$ is a Hurwitz matrix, then the matrix $A_L = A - LC$ where $L = P^{-1}\bar{L}$ has the same spectrum. In order to prove the statement, it suffices to note that

$$\bar{A}_L = PAP^{-1} - (PL)(CP^{-1}) = PA_L P^{-1},$$

i.e., the matrices A_L and $\bar{A}_L$ are similar.

Therefore, without loss of generality of reasoning, we can assume that the pair $\{C, A\}$ is given in the form of Kalman decomposition for the not fully observable system (2.8)

$$A = \begin{pmatrix} A_{11} & 0 \\ A_{21} & A_{22} \end{pmatrix}, \quad C = (C_1; 0), \tag{3.7}$$

where $A_{11} \in \mathbb{R}^{\xi \times \xi}$, $A_{21} \in \mathbb{R}^{(n-\xi)\times\xi}$, $A_{22} \in \mathbb{R}^{(n-\xi)\times(n-\xi)}$, $C_1 \in \mathbb{R}^{l\times\xi}$. A_{11} corresponds to the observable subsystem of dimension $\xi = \operatorname{rank} N(C, A)$, i.e., the pair $\{C_1, A_{11}\}$ is observable.

By virtue of Theorem 2.17, for the reconstructible pair $\{C, A\}$ the spectrum of the nonobservable part, i.e., the matrix A_{22}, lies in $\mathbb{C}_-$ (the left-hand open half-plane of $\mathbb{C}$).

Since the pair $\{C_1, A_{11}\}$ is observable, there exists for it a matrix $L_1 \in \mathbb{R}^{\xi\times l}$ such that $A_{11}^L = A_{11} - L_1 C_1$ possesses the defined spectrum, in particular, this matrix can be made a Hurwitz matrix by a requisite choice of the matrix L_1. Consider a matrix

$$L = \begin{pmatrix} L_1 \\ 0 \end{pmatrix} \in \mathbb{R}^{n\times l}.$$

For such a feedback matrix we have

$$A_L = \begin{pmatrix} A_{11} - L_1 C_1 & 0 \\ A_{21} & A_{22} \end{pmatrix} = \begin{pmatrix} A_{11}^L & 0 \\ A_{21} & A_{22} \end{pmatrix}.$$

Since the diagonal matrices A_{11}^L (for a corresponding choice of L_1) and A_{22} are Hurwitz matrices A_L is also a Hurwitz matrix. The theorem is proved. □

Remark 3.8. Suppose that the pair $\{C, A\}$ is not fully observable, $\lambda_1, \ldots, \lambda_{n-\xi} \in \mathbb{C}$ are the values on which the rank of the Rosenbrock observability matrix is degenerate

$$R(C, A) = \begin{pmatrix} \lambda I - A \\ C \end{pmatrix}$$

(and if the rank of the matrix $R(C, A)$ diminishes by k on a certain λ, then this number enters the set k times, i.e., has multiplicity k). Then the characteristic polynomial of the matrix A_L has the form

$$\det(sI - A_L) = \alpha(s) = \left(\prod_{i=1}^{n-\xi}(s - \lambda_i)\right)\alpha_\xi(s) = \alpha_{n-\xi}^*(s)\alpha_\xi(s),$$

$\alpha_{n-\xi}^*(s)$ is an invariable polynomial whose roots are defined by the properties of C and A, the roots of the polynomial $\alpha_\xi(s)$ of degree ξ may be defined arbitrarily by the requisite choice of the matrix L.

Proof. As was pointed out in the proof of the theorem, the investigation of the spectrum of the matrix A_L can be carried out in any basis. Let us return to the pair $\{C, A\}$ defined in the form (3.7). Suppose that the matrix L is written in the block form

$$L = \begin{pmatrix} L_1 \\ L_2 \end{pmatrix}, \quad L_1 \in \mathbb{R}^{\xi \times l}, \quad L_2 \in \mathbb{R}^{(n-\xi)\times l}.$$

Then

$$A_L = A - LC = \begin{pmatrix} A_{11} - L_1 C_1 & 0 \\ A_{21} - L_2 C_1 & A_{22} \end{pmatrix}.$$

Thus, the spectrum of the matrix A_L consists of two parts: the spectrum of A_{22} and that of $(A_{11} - L_1 C_1)$. Since the pair $\{C, A\}$ is not fully observable, we have a relation $\operatorname{spec}\{A_{22}\} = \{\lambda_1, \dots, \lambda_{n-\xi}\}$.

The pair $\{C_1, A_{11}\}$ is observable, and therefore the spectrum of $(A_{11} - L_1 C_1)$ is defined arbitrarily by the requisite choice of L_1 (and, consequently, of the matrix L). The statement is proved. □

Thus, in the case where the pair $\{C, A\}$ is reconstructible, for system (3.1) we can also use an asymptotic observer of type (3.5), where the feedback matrix L is chosen in accordance with the condition that $A_L = A - LC$ is a Hurwitz matrix. The difference from the case of full observability is that now a part of the spectrum of A_L does not change (and is stable) and the other part is arbitrary. The rate of convergence of the error $e(t) = x(t) - \tilde{x}(t)$ cannot now be arbitrary, it is defined by the unchangeable part of the spectrum of A_L (i.e., by the spectrum of the nonobservable subsystem).

Similar results can be formulated and proved for a control problem. We shall only formulate the main results [1].

Theorem 3.9. *Let the pair $\{A, B\}$ be controllable. Then, for any polynomial $\varphi_n(s) = s^n + \gamma_n s^{n-1} + \dots + \gamma_1$ with real coefficients γ_i, $s \in \mathbb{C}$, there exists a matrix $K \in \mathbb{R}^{m\times n}$ such that $\varphi_n(s)$ is a characteristic polynomial of the matrix $A_K = A - BK$, i.e., $\det(sI - A_K) = \varphi_n(s)$.*

The proof of Theorem 3.9 is similar to that of Theorem 3.1. In this case, for stabilization of system (3.1) with respect to the full-phase vector (i.e., if the whole vector $x(t)$ is known), we can use a linear feedback of the form

$$u(t) = -Kx(t) \tag{3.8}$$

with constant feedback matrix K. System (3.1) closed by a feedback of this kind satisfies the equation

$$\dot{x} = Ax - BKx = A_K x,$$

and, if the matrix K is chosen from the condition that A_K is a Hurwitz matrix, then the transition process $x(t) \to 0$ as $t \to \infty$. Moreover, $x \to 0$ exponentially and the rate of convergence on the whole is defined by the choice of the matrix K.

Now if the pair $\{A, B\}$ is only stabilizable, then we have the following statement.

Theorem 3.10. *Let the pair* $\{A, B\}$ *is stabilizable. Then there exists a matrix* $K \in \mathbb{R}^{m\times n}$ *such that* $A_K = A - BK$ *is a Hurwitz matrix.*

Remark 3.11. The spectrum of the matrix A_K consists of two parts. The part $\lambda_1, \dots, \lambda_{n-\eta} \in \mathbb{C}$ of the spectrum does not depend on the choice of the matrix K and corresponds to the uncontrollable part of the system, i.e., forms the spectrum of the matrix A_{22} from the Kalman decomposition (2.9) for noncompletely controllable systems. This part of the spectrum is stable since the pair $\{A, B\}$ is stable. The remaining part of the spectrum of A_K is defined arbitrarily by the choice of the matrix K. In this case, as before, the feedback (3.8) stabilizes system (3.1) asymptotically (exponentially), but now the rate of convergence is defined by the unchangeable part of the spectrum of the matrix A_K and cannot be defined arbitrarily.

Consider system (3.1). The use of the full-dimensional observer (3.5) and the linear feedback (3.8) allows us to solve the problem of stabilization of system (3.1) with respect to the output (i.e., under the constraint $u = u(y(t))$).

The *principle of separability* of problems of observation and stabilization is valid for the linear stationary system (3.1).

Theorem 3.12. *Suppose that the pair* $\{C, A\}$ *is observable and observer* (3.5) *gives an asymptotic estimate* $\tilde{x}(t)$ *to the unknown phase vector* $x(t)$. *Let the pair* $\{A, B\}$ *be controllable and the feedback with respect to the full-phase vector* (3.8) *stabilizes system* (3.1) *at zero exponentially. Then the feedback with respect to the estimate* $\tilde{x}(t)$

$$u = -K\tilde{x}(t) \tag{3.9}$$

stabilizes system (3.1) *at zero exponentially.*

Proof. It suffices to consider jointly the equations of system (3.1) closed by feedback (3.9) and equations (3.6) for the observation error $e(t) = x(t) - \tilde{x}(t)$. Control (3.9) can be represented as $u = -K(x(t) - e(t))$. Then we have a system of equations

$$\begin{cases} \dot{x} = Ax - BK(x - e) = A_K x + BKe \\ \dot{e} = A_L e. \end{cases} \tag{3.10}$$

The matrix of system (3.10) has a block-diagonal form

$$\tilde{A} = \begin{pmatrix} A_K & BK \\ 0 & A_L \end{pmatrix}$$

with Hurwitz matrices A_K and A_L on the diagonal. Consequently, A is also a Hurwitz matrix and $x(t), e(t) \to 0$ as $t \to \infty$ exponentially. The theorem is proved. □

Remark 3.13. In the case of observability of the pair $\{C, A\}$ and controllability of the pair $\{A, B\}$ the spectrum of the matrices A_L and A_K is defined arbitrarily and, consequently, the degree of stability of the matrix $\tilde{A}$ can also be defined arbitrarily. Thus, if system (3.1) is in the general position, then the rate of convergence to zero of the phase vector of system (3.1) which is closed by feedback (3.9) can be defined arbitrarily.

Now if the pair $\{C, A\}$ is only reconstructible (or the pair $\{A, B\}$ is only stabilizable), then the statement of the theorem remains valid but the rate of convergence of the phase vector of the system for $u(t)$ from (3.9) is defined by the spectrum of nonobservable (or uncontrollable) part.

3.2 Lowered order Luenberger observers

In the automatic control theory a requirement is often advanced to the dimension of the observer which must be lowered to the minimal one. We have described above the scheme of construction of full-dimensional observers for linear stationary systems without uncertainty (i.e., observers whose dimension coincides with the dimension of the system itself). However, observers of lowered order can be constructed for systems of this kind. For the first time, methods of construction of observers of this kind were proposed in works of Luenberger (see [87]).

In what follows we again consider system (3.1) and assume that the pair $\{C, A\}$ is observable.

The latter assumption does not lead to the loss of generality of the arguments. If the pair $\{C, A\}$ is only reconstructible, then system (3.1) can be reduced to the Kalman decomposition (2.8)

$$\left\{\begin{array}{l} \left\{\begin{array}{l} \dot{x}^1 = A_{11}x^1 + B_1 u \\ \dot{x}^2 = A_{21}x^1 + A_{22}x^2 + B_2 u \end{array}\right. \\ \quad y = C_1 x^1, \end{array}\right.$$

where the pair $\{C_1, A_{11}\}$ is observable. If an observer which gives estimate $\tilde{x}^1$ is constructed for the first subsystem, then the observer for the second subsystem has the form

$$\dot{\tilde{x}}^2 = A_{21}\tilde{x}^1 + A_{22}\tilde{x}^2 + B_2 u. \tag{3.11}$$

The observation error $e_2 = x^2 - \tilde{x}^2$ satisfies the equation

$$\dot{e}^2 = A_{22}e^2 + A_{21}e^1,$$

whence, since $e^1 = x^1 - \tilde{x}^1 \to 0$ and A_{22} is a Hurwitz matrix, it follows that $e^2 \to 0$. The observation order of (3.11) is equal to $(n - \xi)$ and cannot be changed. Thus, the order of the observer for the system as a whole depends on the order of the observer for the first full-observable subsystem.

We begin with considering the case of a scalar full-observable system, i.e., $l = 1$. In this case the system can be written in one of the canonical forms, the output $y(t)$ being one of the coordinates of the phase vector, i.e., $(n-1)$ coordinates are unknown. We shall show that under these conditions we can construct an observer of order $(n-1)$.

We shall describe one of the methods of constructing such an observer following [1]. Let the pair $\{C, A\}$ be given in the second canonical form

$$A = \begin{pmatrix} 0 & 0 & \dots & 0 & -\alpha_1 \\ 1 & 0 & \dots & 0 & -\alpha_2 \\ 0 & 1 & \dots & 0 & -\alpha_3 \\ \dots & \dots & \dots & \dots & \dots \\ 0 & 0 & \dots & 1 & -\alpha_n \end{pmatrix}, \qquad C = (0, \dots, 0, 1).$$

In this basis $x_n = y$ is a known coordinate. Let us consider the equations for the first coordinates of system (3.1)

$$\dot{x}' = A'x' + a'y + B'u, \tag{3.12}$$

where $x' = (x_1, \dots, x_{n-1})$, $A' \in \mathbb{R}^{(n-1)\times(n-1)}$ is the principal minor of order $(n-1)$ of the matrix A, $a' \in \mathbb{R}^{(n-1)\times 1}$ is the last column of the matrix A without a_{nn}, $B' \in \mathbb{R}^{(n-1)\times m}$ is a matrix consisting of the first $(n-1)$ rows of the matrix B. We shall construct an observer for x' in the form

$$\dot{\tilde{x}}' = A'\tilde{x}' + a'y + B'u. \tag{3.13}$$

Then the observation error $e' = x' - \tilde{x}'$ satisfies the equation

$$\dot{e}' = A'e',$$

and $e' \to 0$ exponentially if A' is a Hurwitz matrix.

For the matrix A given in the canonical basis we have

$$A' = \begin{pmatrix} 0 & 0 & \dots & 0 \\ 1 & 0 & \dots & 0 \\ 0 & 1 & \dots & 0 \\ \vdots & \vdots & \ddots & \vdots \\ 0 & 0 & \dots & 1 \end{pmatrix},$$

$\alpha(s) = \det(sI - A) = s^n$ not being a Hurwitz polynomial.

However, we can indicate a transformation of the first $(n-1)$ coordinates after which the matrix A' will have any preassigned spectrum. In this case, the indicated transformation does not change the vector C, and, consequently, in the new basis observer (3.13) solves the problem exponentially.

We shall show this transformation in explicit form. Suppose that we are given a polynomial of order $(n-1)$ with real coefficients

$$\varphi_{n-1}(s) = s^{n-1} + \gamma_{n-1}s^{n-2} + \cdots + \gamma_1.$$

Consider a transformation matrix of the form

$$P = \begin{pmatrix} 1 & 0 & \dots & 0 & -\gamma_1 \\ 0 & 1 & \dots & 0 & -\gamma_2 \\ \dots & \dots & \dots & \dots & \dots \\ 0 & 0 & \dots & 1 & -\gamma_{n-1} \\ 0 & 0 & \dots & 0 & 1 \end{pmatrix}, \quad P^{-1} = \begin{pmatrix} 1 & 0 & \dots & 0 & \gamma_1 \\ 0 & 1 & \dots & 0 & \gamma_2 \\ \dots & \dots & \dots & \dots & \dots \\ 0 & 0 & \dots & 1 & \gamma_{n-1} \\ 0 & 0 & \dots & 0 & 1 \end{pmatrix}.$$

Obviously, this transformation does not change $x_n = y$ and, consequently, the vector $C = (0, \dots, 0, 1)$.

Upon the indicated transformation the matrix A assumes the form

$$PAP^{-1} = \left(\begin{array}{ccccc|c} 0 & 0 & \dots & 0 & -\gamma_1 & \{(\alpha_n - \gamma_{n-1})\gamma_1 - \alpha_1\} \\ 1 & 0 & \dots & 0 & -\gamma_2 & \{(\alpha_n - \gamma_{n-1})\gamma_2 - \alpha_2 + \gamma_1\} \\ \dots & \dots & \dots & \dots & \dots & \dots \\ 0 & 0 & \dots & 1 & -\gamma_{n-1} & \{(\alpha_n - \gamma_{n-1})\gamma_{n-1} - \alpha_{n-1} + \gamma_{n-2}\} \\ \hline 0 & 0 & \dots & 0 & 1 & \{\alpha_n + \gamma_n\} \end{array}\right)$$

$$= \left(\begin{array}{c|c} A' & a' \\ \hline 0 \dots 0\, 1 & (\alpha_n + \gamma_n) \end{array}\right),$$

and the characteristic polynomial of the matrix A' assumes the form

$$\det(sI - A') = \varphi_{n-1}(s).$$

If we choose $\varphi_{n-1}(s)$ to be a Hurwitz polynomial, then observer (3.13) of order $(n-1)$ reconstructs the unknown coordinates of the phase vector.

For the phase vector written in the original canonical basis the following relation gives an estimate

$$\hat{x} = P^{-1}\begin{pmatrix} \tilde{x}' \\ y \end{pmatrix}.$$

Note that if the pair $\{C, A\}$ is reconstructible, then, for its observable part, we can construct an observer of order $(\xi - 1)$. Together with observer (3.11) of order $(n - \xi)$ they form an observer of a full-phase vector of order $(n - \xi) + (\xi - 1) = n - 1$, i.e., of the same order as the Luenberger observer for the observable pair $\{C, A\}$. The difference consists in the fact that the rate of convergence in the case of the reconstructible pair $\{C, A\}$ depends on the spectrum of the nonobservable part of the system.

The proposed method of constructing observers of a lowered order can be generalized to the case $l > 1$. Let us consider a system written in the canonical Luenberger form

$$\begin{cases} \dot{x}_i = A_{ii}x_i + \sum\limits_{j=1, j\neq i}^{l} \bar{a}_{ij}y_j + B_i u, \quad i = 1, \dots, l \\ y_i = C_i x_i. \end{cases} \tag{3.15}$$

We denote by

$$\bar{u}_i = \begin{pmatrix} u \\ y_1 \\ \vdots \\ y_{i-1} \\ y_{i+1} \\ \vdots \\ y_l \end{pmatrix}$$

the new known input. Then we can write the subsystems from (3.15) in the form

$$\begin{cases} \dot{x}_i = A_{ii}x_i + \bar{B}_i\bar{u}_i, \quad i = 1, \dots, l \\ y_i = C_i x_i, \end{cases} \tag{3.16}$$

where $\bar{B}_i$ is the known matrix defined by B_i and $\bar{a}_{ij}$, $j = 1, \dots, l$, $j \neq i$.

Each of the subsystems (3.16) can be regarded as an independent system of order ν_i with the known input $\bar{u}_i$ and the known scalar output y_i. We can construct for it an observer of order $\nu_i - 1$. A set of observers of this kind forms an observer of a full-phase vector. The order of such an observer is equal to

$$\sum_{i=1}^{l}(\nu_i - 1) = \left[\sum_{i=1}^{l}\nu_i\right] - l = n - l.$$

Just as in the scalar case, we can also construct an observer of order $(n - l)$ for $l > 1$ for the reconstructible pair.

Thus we have the following statement [1, 87].

Theorem 3.14. *Let the pair $\{C, A\}$ be observable (reconstructible). Then we can construct for system (3.1) an observer of order $(n - l)$ with a defined rate of convergence (the rate of convergence is defined by the nonobservable part of the system).*

The following approach proposed by Luenberger is often used for constructing an observer of a defined order p (in particular, of a minimal order $p = n - l$).

As before, we consider a linear stationary system (3.1)

$$\begin{cases} \dot{x} = Ax + Bu \\ y = Cx. \end{cases}$$

We choose a matrix $F \in \mathbb{R}^{p\times n}$ such that

$$\operatorname{rank}\begin{pmatrix} F \\ C \end{pmatrix} = n. \tag{3.17}$$

In this case, obviously, $p \geq n - l$.

We denote $z = Fx$, and then

$$\begin{pmatrix} z \\ y \end{pmatrix} = \begin{pmatrix} F \\ C \end{pmatrix} x,$$

and, consequently, since the rank of the matrix $\left(\begin{smallmatrix} F \\ C \end{smallmatrix}\right)$ is full, for reconstructing the vector x it suffices to construct an estimate for $z(t)$. In what follows, we assume, for simplicity, that $p = n - l$. In this case we have

$$x = \begin{pmatrix} F \\ C \end{pmatrix}^{-1} \begin{pmatrix} z \\ y \end{pmatrix}.$$

In order to estimate $z(t)$ we can use a Luenberger observer of the form

$$\dot{\tilde{z}} = D\tilde{z} + Ey + Gu, \tag{3.18}$$

where $\tilde{z} \in \mathbb{R}^{n-l}$, D, E, and G are constant matrices, parameters of the observer which must be defined.

Let us consider the observation error $e = \tilde{z} - z$. This error satisfies the equation

$$\dot{e} = De + (DF + EC - FA)x + (G - FB)u. \tag{3.19}$$

The parameters of the observer are chosen from the conditions

$$\begin{cases} G = FB \\ EC = FA - DF \\ D \text{ is a Hurwitz matrix.} \end{cases} \tag{3.20}$$

If conditions (3.20) are fulfilled, the estimation error $e(t)$ satisfies the equation

$$\dot{e} = De,$$

and, since D is a Hurwitz matrix, $e(t) \to 0$ as $t \to \infty$. The relation

$$\tilde{x}(t) = \begin{pmatrix} F \\ C \end{pmatrix}^{-1} \begin{pmatrix} \tilde{z} \\ y \end{pmatrix}$$

gives an estimate for the unknown phase vector $x(t)$.

If system (3.1) is fully observable, then the system of matrix equations (3.17), (3.20) is solvable [76, 87] (i.e., the matrices D, E, G, and F can be determined from it), and the matrix D can have any preassigned symmetric (i.e., complex-conjugate) spectrum. If the pair $\{C, A\}$ is only reconstructible, then the spectrum of D contains an unchangeable part which corresponds to the nonobservable part of system (3.1).

The synthesis of an observer can be carried out, for instance, with the use of the following arguments. Suppose that P is a matrix of the left eigenvectors of the matrix $A_L = A - LC$, the spectrum of A_L being a Hurwitz spectrum, real, and different. Then the matrix P satisfies the equation

$$P(A - LC) = \Lambda P,$$

where $\Lambda = \operatorname{diag}(\lambda_1, \ldots, \lambda_n)$. It is obvious that rank $P = n$, and among its rows we can choose $(n - l)$ as F so that

$$F(A - LC) = \Lambda_F F,$$

and rank $\binom{F}{C} = n$. Comparing this relation with (3.20), we find that $D = \Lambda_F$, $E = FL$.

Conclusion

In Chapter 3 we considered a problem of synthesizing the observers of a phase vector for linear stationary fully determined systems. We have described the algorithms of synthesis of full-dimensional observers for scalar ($l = 1$) and vector ($l > 1$) systems. We have also described Luenberger observers of a lowered order $(n - l)$.

Chapter 4

Functional observers for fully determined linear systems

4.1 Problem statement. Luenberger type functional observers

We shall again consider an n-dimensional linear stationary fully determined observable systems with an l-dimensional output

$$\begin{cases} \dot{x} = Ax + Bu \\ y = Cx. \end{cases} \tag{4.1}$$

We assume everywhere in the sequel that the matrix C is of full rank, i.e., rank $C = l$. In this chapter we solve the problem of constructing functional observers, i.e., the problem of reconstructing a linear functional from the unknown phase vector

$$\sigma = Fx, \tag{4.2}$$

where $F \in \mathbb{R}^{p\times n}$ is a known matrix. Such a problem arises, for instance, when we solve the problem of stabilization of system (4.1) at zero. To solve such a problem we can use, for instance, a feedback of the form $u = -Kx$. If we know the estimate of the full-phase vector $\tilde{x}(t)$, then the relation $u = -K\tilde{x}$ solves the stabilization problem. However, in this case it is not obligatory to reconstruct the whole n-dimensional vector $\tilde{x}(t)$, it is only sufficient to construct an estimate of the functional $\sigma(t) = -Kx$, $\sigma(t) \in \mathbb{R}^m$, the dimension of the functional observer (i.e., the observer that forms the estimate $\tilde{\sigma}(t)$) must turn out to be smaller than that of the Luenberger observer.

The problem of constructing a functional observer for system (4.1), (4.2) was considered in detail in the works by O'Reilly [87], who described an observer similar in structure to the Luenberger observer (3.18). The following procedure was proposed for constructing an observer of order k.

Suppose that a decomposition

$$F = PT + VC,$$

holds for the matrix F, where $P \in \mathbb{R}^{p\times k}$, $T \in \mathbb{R}^{k\times n}$, $V \in \mathbb{R}^{p\times l}$. Then

$$\sigma = Fx = PTx + VCx = Pz + Vy,$$

where $z(t) \in \mathbb{R}^k$ is an unknown vector to be estimated. For its reconstruction we use an observer of the Luenberger type

$$\dot{\tilde{z}} = D\tilde{z} + Ey + Gu. \tag{4.3}$$

Then the estimate for the required functional $\sigma(t)$ is given by

$$\tilde{\sigma}(t) = P\tilde{z} + Vy. \tag{4.4}$$

For the estimate $\tilde{\sigma}(t)$ to converge to $\sigma(t)$ it suffices that $(\tilde{z}(t) - z(t)) \to 0$ as $t \to \infty$. Consider an equation for the error $e = \tilde{z} - z$:

$$\begin{aligned} \dot{e} = \dot{\tilde{z}} - \dot{z} &= D\tilde{z} + Ey + Gu - T(Ax + Bu) \\ &= D(e + Tx) + ECx + Gu - TAx - TBu \\ &= De + (DT + EC - TA)x + (G - TB)u. \end{aligned} \tag{4.5}$$

It [87] the following theorem was proved.

Theorem 4.1. *Observer* (4.3), (4.4) *of order k reconstructs the functional σ from* (4.2) *for system* (4.1) *if and only if the conditions*

$$\begin{aligned} &F = PT + VC, \\ &G = TB, \\ &TA - DT = EC, \\ &D \ \text{ is a Hurwitz matrix,} \end{aligned} \tag{4.6}$$

are fulfilled.

Under these conditions the estimation error $e(t)$ obviously satisfies the equation

$$\dot{e}(t) = De(t),$$

and, since D is a Hurwitz matrix,

$$e(t) \overset{\exp}{\to} 0 \text{ as } t \to \infty.$$

The general scheme of the synthesis of the observer can be represented, for instance, as follows. Let Θ be a matrix of the left eigenvectors of the matrix $A_L = A - LC$ corresponding to different real stable eigenvalues which form a diagonal matrix Λ, i.e.,

$$\Theta A_L = \Lambda\Theta.$$

Suppose that T is a collection of k rows from Θ such that rank $\binom{T}{C} = k + l$, and then the matrices P and V can be found from the equation

$$F = (P\ V)\begin{pmatrix} T \\ C \end{pmatrix}.$$

Postmultiplying this equation by the matrix $\begin{pmatrix} T \\ C \end{pmatrix}^\top$, we obtain

$$F \begin{pmatrix} T \\ C \end{pmatrix}^\top = (P\ V) N_k,$$

where $N_k \in \mathbb{R}^{(k+l)\times(k+l)}$. The minimal number k for which the matrix N_k is invertible gives the solution of the problem since

$$F \begin{pmatrix} T \\ C \end{pmatrix}^\top N_k^{-1} = (P V).$$

In this case, $D = \Lambda_T$ is a part of the diagonal matrix Λ, i.e., $TA_L = \Lambda_T T$ and $E = TL$.

Note that $\det N_k = \det(CC^\top - CT^\top(TT^\top)^{-1}TC^\top)$ and the determination of the minimal k is connected with the sorting out of the collections of k rows of the matrix Θ which are linearly independent of the rows of the matrix C. It is obvious that the matrix N_{n-l} is invertible for $k = n - l$.

Since no clear algorithm was proposed in the general case, the following questions arise.

(1) How, for a given k, can we find the matrices P, T, V, G, E, and D satisfying conditions (4.6)?

(2) What is the minimal value of k for which we can construct a functional observer?

In the general case, where $y(t)$ and $\sigma(t)$ are vectors (i.e., $l > 1$, $p > 1$), these questions are difficult for answering. The author of [87] gives results for the cases where $y(t)$ or $\sigma(t)$ are scalar variables. In particular, the following statement exists for the order of the observer.

Theorem 4.2 (Roman and Bullock [93]). *An observer of order k with any preassigned (symmetric) spectrum*

(i) *reconstructs the functional $\sigma = Fx$, $\sigma \in \mathbb{R}^p$, for a system with a scalar output (i.e., for $l = 1$) if and only if $k \geq n - 1$,*

(ii) *reconstructs the scalar functional $\sigma = Fx = \sum_{i=1}^{l} F_i x_i$ (where $x_i \in \mathbb{R}^{\nu_i}$ are parts of the phase vector x in the canonical Luenberger representation* (2.11)*, $F_i \in \mathbb{R}^{1\times\nu_i}$) if and only if $k \geq d - 1$, where d is the length of the maximal nonzero part among the vectors F_i ($d - 1 \leq \nu - 1$, ν is the observability index of the pair $\{C, A\}$, $\nu = \max \nu_i$).*

For $p = 1$ the author of [87] showed the technique of constructing an observer of order $(\nu - 1)$. It follows from Theorem 4.2 for $l = 1$, $p > 1$ that the minimal order of an observer with a completely defined spectrum is equal to $n - 1 = \nu - 1$ and coincides with the dimension of the Luenberger observer for a full-phase vector.

However, in some cases, if we refuse the condition of a completely defined spectrum, we can lower the order of the observer. In recent years this problem was actively investigated. In particular, the author of [122] obtained conditions for the existence of observers of order p which coincides with the dimension of the functional. The authors of [25–29] obtained necessary and sufficient conditions for the synthesis of observers of a defined order. Somewhat later similar conditions were given in paper [123].

The technique of determining the minimal dimension of a functional observer is given in papers [27, 28]. We shall describe it following the indicated papers.

4.2 Reconstruction of scalar functionals

We shall begin with considering the case of a scalar output and a scalar functional, i.e., $p = l = 1$. In this case, as Theorem 4.2 indicates, we can construct a functional observer of order $(n - 1)$. The dimension of this observer (and, consequently, the technique of its construction) coincides with that of the Luenberger observer. Of interest is the possibility of constructing an observer of order $k < n - 1$. Let us investigate this possibility.

Without loss of generality, we assume that the pair $\{C, A\}$ is given in the canonical form

$$A = \begin{pmatrix} 0 & 0 & \dots & 0 & -\alpha_1 \\ 1 & 0 & \dots & 0 & -\alpha_2 \\ \dots & \dots & \dots & \dots & \dots \\ 0 & 0 & \dots & 1 & -\alpha_n \end{pmatrix}, \quad C = (0, 0, \dots, 0, 1), \tag{4.7}$$

where α_i are coefficients of the characteristic polynomial of the matrix A, i.e.,

$$\alpha(s) = \det(sI - A) = s^n + \alpha_n s^{n-1} + \dots + \alpha_1.$$

Suppose that the pair $\{F, A\}$ is observable and the vector F, in the indicated basis, has the form

$$F = (f_1, \dots, f_n).$$

In what follows, without loss of generality, we assume that the vectors C and F are not collinear. In order to solve the problem, we shall use the method of pseudoinputs, namely, consider the system

$$\begin{cases} \dot{x} = Ax + Bu + Lv \\ y = Cx \end{cases} \tag{4.8}$$

with an additional input $v \in \mathbb{R}$ ("pseudoinput"). The form of the vector $L \in \mathbb{R}^{n\times 1}$ will be defined below.

In the case where the pseudoinput $v \equiv 0$ this system coincides with system (4.1). For system (4.8) the transfer functions from the new input v to the output $y(t)$ and to the functional $\sigma(t)$ are defined:

$$W_1(s) = C(sI - A)^{-1}L, \quad y = W_1 v,$$

$$W_2(s) = F(sI - A)^{-1}L, \quad \sigma = W_2 v$$

(in the last expressions, for simplicity, we preserved the same notation for the Laplace transformations for $y(t)$ and $\sigma(t)$ as for the preimages). The functions $W_1(s)$ and $W_2(s)$ are fractional-rational, i.e., are the relation of polynomials

$$W_1(s) = \frac{\beta_1(s)}{\alpha(s)}, \quad W_2(s) = \frac{\beta_2(s)}{\alpha(s)},$$

where $\beta_1(s)$ and $\beta_2(s)$ are polynomials of degree not higher than $(n-1)$. The polynomials $\beta_1(s)$ and $\beta_2(s)$ are defined by the choice of the vector $L = (l_1, l_2, \dots, l_n)^\top$. Moreover, since the matrices A and C are given in the canonical form, it follows that

$$\beta_1(s) = s^{n-1}l_n + s^{n-2}l_{n-1} + \dots + l_1.$$

Suppose that the vector L is chosen such that the polynomials $\beta_1(s)$, $\beta_2(s)$, and $\alpha(s)$ do not have common roots, i.e., the triples $\{C, A, L\}$ and $\{F, A, L\}$ are in the general position. Then the transfer function from the measurable output of the system $y(t)$ to the unknown functional $\sigma(t)$ is defined:

$$\sigma = \frac{\beta_2(s)}{\beta_1(s)} y = \tilde{W}(s)y. \tag{4.9}$$

For constructing an asymptotic observer, it is sufficient that the transfer function $\tilde{W}(s)$ should be physically realizable, i.e., the condition

$$\deg \beta_2(s) \le \deg \beta_1(s) \tag{4.10}$$

should be fulfilled and the denominator $\tilde{W}(s)$, i.e., the polynomial $\beta_1(s)$, should be a Hurwitz polynomial. Moreover, if $\deg \beta_1(s) = k$, then the functional observer which reconstructs $\sigma(t)$ will have an order k. Thus, the problem of constructing a functional observer of order k reduces to the search for a vector L such that condition (4.10) is fulfilled and $\beta_1(s)$ satisfies the conditions

$$\begin{aligned} &\deg \beta_1(s) = k, \\ &\beta_1(s) \text{ is a Hurwitz polynomial.} \end{aligned} \tag{4.11}$$

When conditions (4.10), (4.11) are fulfilled, we can easily indicate an algorithm for constructing an observer of order k. If $\deg \beta_1(s) = k$, then, without loss of generality, we have

$$\beta_1(s) = s^k + l_k s^{k-1} + \dots + l_1,$$

i.e., $L = (l_1, \dots, l_k, 1, 0, \dots, 0)$. Since the polynomials $\beta_1(s)$ and $\alpha(s)$ are uncancellable, we can use a nondegenerate transformation of coordinates in order to reduce system (4.8) to the canonical form of controllability

$$\begin{cases} \dot{x} = \tilde{A}x + \tilde{B}u + \tilde{L}v \\ y = \tilde{C}x \\ \sigma = \tilde{F}x, \end{cases} \tag{4.12}$$

where

$$\tilde{A} = \begin{pmatrix} 0 & 1 & 0 & \dots & 0 & 0 \\ 0 & 0 & 1 & \dots & 0 & 0 \\ \dots & \dots & \dots & \dots & \dots & \dots \\ -\alpha_1 & -\alpha_2 & -\alpha_3 & \dots & -\alpha_{n-1} & -\alpha_n \end{pmatrix}, \quad \tilde{L} = \begin{pmatrix} 0 \\ \vdots \\ 0 \\ 1 \end{pmatrix},$$

$$\tilde{C} = (l_1, \dots, l_k, 1, 0, \dots, 0)$$

(we denote the phase vector x, the inputs u and v, and the outputs y and σ as before). Since the relative order of system (4.12) with respect to the output σ is not lower than with respect to the output y (by virtue of condition (4.10)), it follows that

$$\tilde{F} = (\tilde{f}_1, \dots, \tilde{f}_k, \tilde{f}_{k+1}, 0, \dots, 0).$$

In this case, some of the first $(k+1)$ components $\tilde{f}_i$ of the vector $\tilde{F}$ may be zero. Since $y = l_1 x_1 + \dots + l_k x_k + x_{k+1}$, in the indicated coordinates the unknown functional $\sigma(t)$ has the form

$$\begin{aligned} \sigma &= \sum_{i=1}^{k+1} \tilde{f}_i x_i = \sum_{i=1}^{k} \tilde{f}_i x_i + \tilde{f}_{k+1}(y - l_1 x_1 - \dots - l_k x_k) \\ &= \sum_{i=1}^{k} (\tilde{f}_i - \tilde{f}_{k+1} l_i) x_i + \tilde{f}_{k+1} y. \end{aligned}$$

Thus, in order to reconstruct $\sigma(t)$, it suffices to construct estimates of the first k coordinates of the phase vector x. Let us consider the first k equations of system (4.12):

$$\begin{cases} \dot{x}_1 = x_2 + \tilde{b}_1 u \\ \dot{x}_2 = x_3 + \tilde{b}_2 u \\ \quad \vdots \\ \dot{x}_k = x_{k+1} + \tilde{b}_k u. \end{cases}$$

Taking into account that $x_{k+1} = y - \sum_{i=1}^{k} l_i x_i$, we can write these equations as

$$\begin{cases} \dot{x}_1 = x_2 + \tilde{b}_1 u \\ \dot{x}_2 = x_3 + \tilde{b}_2 u \\ \quad \vdots \\ \dot{x}_{k-1} = x_k + \tilde{b}_{k-1} u \\ \dot{x}_k = -l_1 x_1 - \dots - l_k x_k + \tilde{b}_k u + y. \end{cases} \tag{4.13}$$

Since the input $u(t)$ of the system and its output $y(t)$ are known, in order to reconstruct $x' = (x_1, \dots, x_k)^\top$ we can use the observer

$$\begin{cases} \dot{\tilde{x}}_1 = \tilde{x}_2 + \tilde{b}_1 u \\ \quad \vdots \\ \dot{\tilde{x}}_{k-1} = \tilde{x}_k + \tilde{b}_{k-1} u \\ \dot{\tilde{x}}_k = -l_1 \tilde{x}_1 - \cdots - l_k \tilde{x}_k + \tilde{b}_k u + y. \end{cases} \tag{4.14}$$

The observation error $e' = \tilde{x}' - x'$ satisfies the system

$$\begin{cases} \dot{e}_1 = e_2 \\ \quad \vdots \\ \dot{e}_{k-1} = e_k \\ \dot{e}_k = -l_1 e_1 - \cdots - l_k e_k, \end{cases} \tag{4.15}$$

and, since $\beta_1(s)$ is a Hurwitz polynomial (the characteristic polynomial of system (4.15)), it follows that $e' \to 0$ as $t \to \infty$. The relation

$$\tilde{\sigma}(t) = \sum_{i=1}^{k} (\tilde{f}_i - \tilde{f}_{k+1} l_i) \tilde{x}_i + \tilde{f}_{k+1} y$$

gives an estimate for $\sigma(t)$. Thus, for constructing a functional observer of order k, it is sufficient that there should exist a vector

$$L = (l_1, \dots, l_k, 1, 0, \dots, 0) \tag{4.16}$$

such that conditions (4.10), (4.11) should be fulfilled. Let us find out in what cases such a vector exists.

Suppose that the vector L has form (4.16), $\beta_1(s) = s^k + l_k s^{k-1} + \cdots + l_1$ is a Hurwitz polynomial of order k. In what follows we shall call the vector $L' = (l_1, \dots, l_k)$, or its extended version $L_k^\top = (L', 1)$ which define the Hurwitz polynomial $\beta_1(s)$, also *Hurwitz vectors*. By virtue of condition (4.10), $\deg \beta_2(s) \le k$. This means that the relative order of the transfer function $W_2(s)$ from the pseudoinput v to the unknown output $\sigma(t)$ is not lower than $(n - k)$, i.e., the conditions

$$FL = 0,\ FAL = 0,\ \dots,\ FA^{n-k-2} L = 0 \tag{4.17}$$

are fulfilled. Since $F = (f_1, \dots, f_n)$, we can take into account the explicit form of the matrix A from (4.7) and the vector L from (4.16) and write conditions (4.17) as

$$\begin{cases} f_1 l_1 + f_2 l_2 + \cdots + f_k l_k + f_{k+1} = 0 \\ f_2 l_1 + f_3 l_2 + \cdots + f_{k+1} l_k + f_{k+2} = 0 \\ \qquad \vdots \\ f_{n-k-1} l_1 + f_{n-k} l_2 + \cdots + f_{n-2} l_k + f_{n-1} = 0. \end{cases}$$

The latter system can be written as a linear system with respect to the unknown vector $L' = (l_1, \dots, l_k)^\top$,

$$\begin{pmatrix} f_1 & f_2 & \dots & f_k \\ f_2 & f_3 & \dots & f_{k+1} \\ \dots & \dots & \dots & \dots \\ f_{n-k-1} & f_{n-k} & \dots & f_{n-2} \end{pmatrix} \begin{pmatrix} l_1 \\ l_2 \\ \dots \\ l_k \end{pmatrix} = - \begin{pmatrix} f_{k+1} \\ f_{k+2} \\ \dots \\ f_{n-1} \end{pmatrix}. \tag{4.18}$$

The matrix of this system has a size $(n-k-1) \times k$, the column of the constant terms has a dimension $1 \times (n-k-1)$, the matrix and the column are fully defined by the parameters of the vector F which defines the functional $\sigma = Fx$ in the canonical basis.

Thus, if system (4.1) is observable, has a scalar output (i.e., $l = 1$), and is given in the canonical form (4.7), then the scalar functional $\sigma = Fx$ can be reconstructed by an observer of order k if there exists a Hurwitz vector $L' = (l_1, \dots, l_k)^\top$ satisfying equations (4.18).

Let us now show the necessity of the conditions indicated above. If there exists an observer of order k, then it is described by a system of differential equations

$$\begin{cases} \dot{z} = Dz + Ey + Gu \\ \tilde{\sigma} = Pz + \gamma y, \end{cases} \tag{4.19}$$

where $z \in \mathbb{R}^k$, D, E, G, P, and γ are constant matrices of the corresponding sizes. In this case, for (4.19) the transfer function from y to σ is defined,

$$\tilde{W}(s) = \gamma + P(sI - D)^{-1}E = \frac{\beta_2(s)}{\beta_1(s)},$$

where $\beta_1(s) = \det(sI - D)$ is a Hurwitz polynomial of order k and $\beta_2(s)$ is a polynomial of a degree not higher than k. Then

$$\sigma = \tilde{W}(s)y = \frac{\beta_2(s)}{\beta_1(s)}y.$$

Suppose that, as before, $\alpha(s)$ is a characteristic polynomial of system (4.1). We denote by v the signal satisfying the condition

$$v = \frac{\alpha(s)}{\beta_1(s)}y.$$

Then

$$\sigma = \frac{\beta_2(s)}{\alpha(s)}v, \quad y = \frac{\beta_1(s)}{\alpha(s)}v.$$

These relations imply that there exists a pseudoinput v and a vector $L = (l_1, \dots, l_k, 1, 0, \dots, 0)$ such that the subvector $L' = (l_1, \dots, l_k)$ is a Hurwitz subvector which satisfies system (4.18). Thus we have the following theorem.

Theorem 4.3. *Suppose that in system* (4.1) $l = 1$, *the pairs* $\{C, A\}$ *and* $\{F, A\}$ *are observable, and* rank $\binom{F}{C} = 2$. *The scalar functional* $\sigma = Fx$ *can be reconstructed by an observer of order k if and only if there exists a Hurwitz vector* $L' = (l_1, \dots, l_k)$, *satisfying* (4.18), *where* f_i *are coefficients of the vector* F *in a canonical basis.*

Remark 4.4. System (4.18) has a solution if the rank condition

$$\operatorname{rank}\begin{pmatrix} f_1 & f_2 & \dots & f_k \\ f_2 & f_3 & \dots & f_{k+1} \\ \dots & \dots & \dots & \dots \\ f_{n-k-1} & f_{n-k} & \dots & f_{n-2} \end{pmatrix} = \operatorname{rank}\begin{pmatrix} f_1 & f_2 & \dots & f_{k+1} \\ f_2 & f_3 & \dots & f_{k+2} \\ \dots & \dots & \dots & \dots \\ f_{n-k-1} & f_{n-k} & \dots & f_{n-1} \end{pmatrix} \tag{4.20}$$

is fulfilled, and therefore condition (4.20), which can be easily verified, is necessary for the existence of such an observer.

Let us consider a scalar system with one output ($l = 1$). The condition for existence of an observer of order k for the functional $\sigma = Fx$ in this case has form (4.18), where L is a Hurwitz column.

Let $k = 1, n > 2$. Then (4.18) turns into a system

$$\begin{pmatrix} f_1 \\ f_2 \\ \vdots \\ f_{n-2} \end{pmatrix} (l_1) = -\begin{pmatrix} f_2 \\ f_3 \\ \vdots \\ f_{n-1} \end{pmatrix},$$

where $l_1 > 0$.

This system of linear equations is equivalent to the equations

$$l_1 = -\frac{f_2}{f_1}, \quad f_i = -l_1 f_{i-1}, \quad i = 3, \dots, n-1.$$

Thus, taking into account that $l_1 > 0$, we find that the solvability of the system is equivalent to the conditions

$$\begin{aligned} \frac{f_2}{f_1} &< 0, \\ f_i &= \frac{f_2}{f_1} f_{i-1}, \quad i = 3, \dots, n-1. \end{aligned} \tag{4.18$'$}$$

Corollary 4.5. *Conditions* (4.18′) *are necessary and sufficient for the existence of a first order observer.*

Suppose now that $k = 2, n > 5$. Conditions (4.18) assume the form

$$\begin{pmatrix} f_1 & f_2 \\ f_2 & f_3 \\ \vdots & \vdots \\ f_{n-3} & f_{n-2} \end{pmatrix} \begin{pmatrix} l_1 \\ l_2 \end{pmatrix} = -\begin{pmatrix} f_3 \\ f_4 \\ \vdots \\ f_{n-1} \end{pmatrix},$$

where $l_1 > 0$ and $l_2 > 0$. It is easy to see that if $f_1 f_3 - f_2^2 \neq 0$, then

$$\begin{pmatrix} l_1 \\ l_2 \end{pmatrix} = -\begin{pmatrix} f_1 & f_2 \\ f_2 & f_3 \end{pmatrix}^{-1} \begin{pmatrix} f_3 \\ f_4 \end{pmatrix}$$

or

$$l_1 = \frac{f_2 f_4 - f_3^2}{f_1 f_3 - f_2^2}, \quad l_2 = \frac{f_2 f_3 - f_4 f_1}{f_1 f_3 - f_2^2}.$$

Thus, the solvability of (4.18) is equivalent to

$$\begin{gathered} \frac{f_2 f_4 - f_3^2}{f_1 f_3 - f_2^2} > 0, \quad \frac{f_2 f_3 - f_4 f_1}{f_1 f_3 - f_2^2} > 0, \quad f_1 f_3 - f_2^2 \neq 0, \\ f_i = f_{i-2} \frac{f_2 f_4 - f_3^2}{f_1 f_3 - f_2^2} + f_{i-1} \frac{f_2 f_3 - f_4 f_1}{f_1 f_3 - f_2^2}, \quad i = 5, \ldots, n-1. \end{gathered} \tag{4.18''}$$

We have the following corollary.

Corollary 4.6. *Conditions* (4.18″) *are necessary and sufficient for the existence of a second-order observer.*

Using Theorem 4.3, we can give an algorithm for constructing a functional observer of a minimal order.

1. Find a minimal value $k = k^*$ for which condition (4.20) is fulfilled. For any $k \geq k^*$ system (4.18) is solvable.

2. By means of search by exhaustion of values of k ($k^* \leq k \leq n-1$) find the minimal value $k = k^{**}$ for which there is a Hurwitz column L' among the solutions of system (4.18).

3. For the found k^{**} and L' construct an observer of the minimal order k^{**}.

Let us consider some examples of realization of this algorithm.

Example 4.7. Consider a third-order system ($n = 3$)

$$\begin{cases} \dot{x} = Ax \\ y = Cx, \end{cases}$$

where

$$A = \begin{pmatrix} 0 & 0 & -2 \\ 1 & 0 & 1 \\ 0 & 1 & 5 \end{pmatrix}, \quad C = (0, 0, 1).$$

It is easy to see that the pair $\{C, A\}$ is observable and is given in a canonical form. Suppose that we have to reconstruct a functional

$$\sigma(t) = x_1 - 2x_2 + 6x_3.$$

In this case $F = (1 \quad -2 \quad 6)$. We begin with finding the minimal k^* for which condition (4.20) is fulfilled.

For $k = 1$ this condition assumes the form

$$\operatorname{rank}(1) = \operatorname{rank}(1 \quad -2),$$

and, consequently, $k^* = 1$. Let us verify whether there is a Hurwitz column among the solutions of system (4.18) for $k = 1$. For $k = 1$, system (4.18) has the form

$$(1)l_1 = (2).$$

The solution of this equation $L' = (l_1) = (2)$ corresponds to the Hurwitz polynomial $\beta(s) = s + 2$, i.e., L' is a Hurwitz column. Consequently, we can construct for $\sigma(t)$ a first-order observer. The observer

$$\begin{cases} \dot{z} = -2z + 24y \\ \tilde{\sigma} = z + 2y \end{cases}$$

gives an estimate for $\sigma(t)$.

Example 4.8. Consider a system of order 4, where

$$A = \begin{pmatrix} 0 & 0 & 0 & 1 \\ 1 & 0 & 0 & -1 \\ 0 & 1 & 0 & 3 \\ 0 & 0 & 1 & 2 \end{pmatrix}, \qquad C = (0, 0, 0, 1).$$

The pair $\{C, A\}$ is again observable and given in a canonical form.

Let us consider a problem of reconstruction of the functional

$$\sigma(t) = x_1 + 2x_2 + 3x_3 + 4x_4.$$

Let us find k^*, which is the minimal k for which condition (4.20) is fulfilled.

For $k = 1$ we have

$$\operatorname{rank}\begin{pmatrix} 1 \\ 2 \end{pmatrix} \neq \operatorname{rank}\begin{pmatrix} 1 & -2 \\ 2 & -3 \end{pmatrix},$$

i.e., condition (4.20) is not fulfilled.

For $k = 2$

$$\operatorname{rank}(1 \quad 2) = \operatorname{rank}(1 \quad 2 \quad -3),$$

i.e., condition (4.20) is fulfilled, $k^* = 2$.

Let us verify whether there is a Hurwitz column for $k = 2$ among the solutions of system (4.18). For $k = 2$, system (4.18) has the form

$$(1 \quad 2)\begin{pmatrix} l_1 \\ l_2 \end{pmatrix} = (-3).$$

The general solution of this equation is

$$l_1 = -2l_2 - 3, \quad l_2 \in \mathbb{R}.$$

This solution is associated with the polynomial $s^2 + l_1 s + (-2l_1 - 3)$ which is a Hurwitz polynomial if and only if

$$\begin{cases} l_1 > 0 \\ -2l_1 - 3 > 0 \end{cases} \Rightarrow \begin{cases} l_1 > 0 \\ l_1 < -\dfrac{3}{2}. \end{cases}$$

This system of inequalities has no solutions and, consequently, the minimal dimension of the observer which reconstructs the given functional $\sigma(t)$ is equal to $k = 3 = (n-1)$, i.e., coincides with the dimension of the Luenberger observer which reconstructs the full-phase vector of the system.

Note that in the given example all coefficients of the vector $f_i > 0$. For this case, proceeding from Theorem 4.3, we can obtain a simple corollary.

Corollary 4.9. *Suppose that the conditions of Theorem* 4.3 *are fulfilled for system* (4.1) *and functional* $\sigma = Fx$. *Suppose, moreover, that* $f_i > 0$ *for all* $i = 1, \ldots, n-1$. *Then the functional* $\sigma(t) = Fx$ *cannot be reconstructed by an observer of an order lower than* $(n-1)$ (*i.e., its reconstruction is possible only by an observer of the maximal order*).

Proof. Consider equation (4.8) under the indicated conditions imposed on the functional. Since $f_i > 0$, all coefficients of the matrix

$$H_k = \begin{pmatrix} f_1 & f_2 & \cdots & f_k \\ f_2 & f_3 & \cdots & f_{k+1} \\ \cdots & \cdots & \cdots & \cdots \\ f_{n-k-1} & f_{n-k} & \cdots & f_{n-2} \end{pmatrix}$$

as well as coefficients of the column

$$h_k = \begin{pmatrix} f_{k+1} \\ f_{k+2} \\ \cdots \\ f_{n-1} \end{pmatrix}$$

are strictly positive for any $k < n-1$. Consequently, any solution $L'_k = (l_1, \ldots, l_k)^\top$ of the equation

$$H_k L'_k = -h_k$$

must contain negative elements l_i. Therefore the column L' cannot be a Hurwitz column for any $k < n-1$. The statement is proved. □

Let us consider one more example.

Example 4.10. Suppose that a system of order four is given in a canonical form, and the matrices A and C have the form

$$A = \begin{pmatrix} 0 & 0 & 0 & -2 \\ 1 & 0 & 0 & -3 \\ 0 & 1 & 0 & 2 \\ 0 & 0 & 1 & -1 \end{pmatrix}, \qquad C = (0, 0, 0, 1).$$

Suppose that we have to reconstruct the functional $\sigma(t) = Fx$, where

$$F = (1, 1, -5, 3).$$

Let us find the minimal value of k using the algorithm proposed above.

If $k = 1$, then condition (4.20) is not fulfilled since

$$\operatorname{rank}\begin{pmatrix} 1 \\ 1 \end{pmatrix} \neq \operatorname{rank}\begin{pmatrix} 1 & 1 \\ 1 & -5 \end{pmatrix}.$$

For $k = 2$ we have

$$\operatorname{rank}(1 \ \ 1) = \operatorname{rank}(1 \ \ 1 \ \ -5),$$

and therefore $k^* = 2$. Let us consider equations (4.18) for $k = 2$:

$$(1 \ \ 1)\begin{pmatrix} l_1 \\ l_2 \end{pmatrix} = (5).$$

The general solution of this system is $l_2 = 5 - l_1$. Among these solutions there are Hurwitz columns $(l_1, l_2)^\top$ when $l_1 > 0$, $l_2 > 0$. In particular, this condition is satisfied by the column $L' = \binom{2}{3}$ corresponding to the Hurwitz polynomial $\beta(s) = s^2 + 3s + 2$ with roots $\lambda_1 = -1$, $\lambda_2 = -2$.

The observer for $\sigma(t)$ can be constructed as

$$\begin{cases} \dot{z}_1 = -z_1 + 3y \\ \dot{z}_2 = -2z_2 + 4y \\ \tilde{\sigma}(t) = 3z_1 - 2z_2 - 13y. \end{cases}$$

In this case $k^{**} = k^* = 2$.

We can see from these examples that for the scalar case ($l = p = 1$) different situations can be realized, in particular, certain functionals can be reconstructed by scalar observers and some of them only by Luenberger observers for a full-phase vector, and some other by observers of order k, where $l < k < n - 1$.

4.3 Reconstruction of vector functionals

The proposed approach to the construction of functional observers can be generalized to the case of the vector functional $\sigma(t) = Fx$, where $F \in \mathbb{R}^{p\times n}$, $p > 1$. Suppose that the rows of the matrix F have the form

$$F^i = (f_1^i, f_2^i, \dots, f_n^i), \quad i = 1, \dots, p. \tag{4.21}$$

Each row is associated with the scalar functional $\sigma_i = F^i x$, $i = 1, \dots, p$. In order to solve the problem we again use the method of pseudoinputs. Consider system (4.8). For the given vector $L \in \mathbb{R}^{n\times 1}$ the transfer functions from the pseudoinput v to the output y will be defined as well as to every scalar functional $\sigma_i(t)$:

$$y = W^*(s)v, \qquad W^*(s) = C(sI - A)^{-1}L = \frac{\beta^*(s)}{\alpha(s)};$$

$$\sigma_i = W_i(s)v, \qquad W_i(s) = F^i(sI - A)^{-1}L = \frac{\beta_i(s)}{\alpha(s)}.$$

Consequently, the transfer functions from the output y to the functionals σ_i are defined:

$$\sigma_i = \tilde{W}_i(s)y, \quad \tilde{W}_i(s) = \frac{\beta_i(s)}{\beta^*(s)}.$$

In order to construct an observer of order k for all σ_i (i.e., for $\sigma(t) \in \mathbb{R}^p$) it is sufficient that the polynomial $\beta^*(s)$ should have an order k, should be a Hurwitz polynomial, and the degrees of the polynomials $\beta_i(s)$ should not exceed k. In this case, the algorithm of constructing an observer is similar to the algorithm for the scalar case.

By analogy with the scalar case we have the following theorem.

Theorem 4.11. *Suppose that system* (4.1) *is observable,* $l = 1$, *the pairs* $\{C, A\}$ *and* $\{F, A\}$ *are observable, the functional* $\sigma = Fx$, *where* $F \in \mathbb{R}^{p\times n}$, $\operatorname{rank}\binom{F}{C} = p + 1$, *has form* (4.21) *in the canonical basis. This functional can be reconstructed by an observer of order* k *if and only if there exists a Hurwitz column* $L' = (l_1, \dots, l_k)^\top$ *among the solutions of the linear system*

$$\begin{pmatrix} f_1^1 & f_2^1 & \cdots & f_k^1 \\ f_2^1 & f_3^1 & \cdots & f_{k+1}^1 \\ \cdots & \cdots & \cdots & \cdots \\ f_{n-k-1}^1 & f_{n-k}^1 & \cdots & f_{n-2}^1 \\ f_1^2 & f_2^2 & \cdots & f_k^2 \\ f_2^2 & f_3^2 & \cdots & f_{k+1}^2 \\ \cdots & \cdots & \cdots & \cdots \\ f_{n-k-1}^2 & f_{n-k}^2 & \cdots & f_{n-2}^2 \\ f_1^3 & f_2^3 & \cdots & f_k^3 \\ \cdots & \cdots & \cdots & \cdots \\ f_{n-k-1}^p & f_{n-k}^p & \cdots & f_{n-2}^p \end{pmatrix} \begin{pmatrix} l_1 \\ l_2 \\ \vdots \\ l_k \end{pmatrix} = - \begin{pmatrix} f_{k+1}^1 \\ f_{k+2}^1 \\ \vdots \\ f_{n-1}^1 \\ f_{k+1}^2 \\ \vdots \\ f_{n-1}^2 \\ f_{k+1}^3 \\ \vdots \\ f_{n-1}^p \end{pmatrix}. \tag{4.22}$$

Remark 4.12. We denote by H_k and h_k the matrix and the column of constant terms of system (4.22) for a given k, i.e.,

$$H_k L' = -h_k.$$

Then the rank condition

$$\operatorname{rank}(H_k) = \operatorname{rank}(H_k, h_k) \tag{4.23}$$

will be a necessary condition for the existence of an observer of order k. The minimal value k^* for which it is fulfilled gives a lower bound for the minimal dimension of the functional observer.

In order to define the minimal dimension of the observer in the case of a vector functional, we can use the procedure, described above, for defining the minimal dimension of the functional observer for a scalar case with a natural replacement of conditions (4.18), (4.20) by conditions (4.22) and (4.23), respectively. In addition, the analog of Corollary 4.9 is also valid in this case, where the positiveness of all f_j^i $(j = 1, \dots, n-1, i = 1, \dots, p)$ must be required.

4.4 Method of scalar observers

In order to solve the problem of constructing a minimal functional observer, we can also use another approach based on scalar observers. This method gives the same result as the method of pseudoinputs but it can be more easily generalized to the case where $l > 1$.

Let us begin with the simplest case where the output of system (4.1) $y(t)$ and the unknown functional $\sigma(t)$ from (4.2) are scalar variables (i.e., $l = 1$, $p = 1$). For this case, we shall find out when the functional $\sigma(t)$ can be reconstructed by a scalar observer.

We shall construct this observer in the form

$$\dot{\tilde{\sigma}} = \lambda\tilde{\sigma} + hu + gy,$$

where the constants λ, h, and g must be further defined. The observation error $\varepsilon = \hat{\sigma} - \sigma$ satisfies the equation

$$\dot{\varepsilon} = \dot{\tilde{\sigma}} - \dot{\sigma} = \lambda\tilde{\sigma} + hu + gy - FAx - FBu.$$

Taking into account that $\tilde{\sigma} = \sigma + \varepsilon = Fx + \varepsilon$ and $y = Cx$, we obtain

$$\dot{\varepsilon} = (-FA + gC + \lambda F)x + \lambda\varepsilon + (h - FB)u,$$

and if the conditions

$$\begin{aligned} &h = FB, \\ &F(\lambda I - A) = -gC, \quad \lambda < 0, \end{aligned} \tag{4.24}$$

are fulfilled, then

$$\dot{\varepsilon} = \lambda\varepsilon,$$

and, consequently, $\varepsilon(t) \to 0$ as $t \to \infty$ exponentially.

The first condition from (4.24) can be fulfilled by means of a choice of h for any defined F and B (therefore, in what follows, we assume for simplicity that $u \equiv 0$ since the influence of the known control can always be compensated). However, not for every F there exists $\lambda < 0$ satisfying (4.24). Let us find out what, for the defined λ and g, is the structure of the vector F satisfying the second equation from (4.24).

We assume, furthermore, that the pair $\{C, A\}$ is given in the canonical form (4.7). Suppose that $F = (f_1, \ldots, f_n)$ in the indicated basis. Then the second equation from (4.24) will assume the form

$$\begin{cases} f_2 = \lambda f_1 \\ f_3 = \lambda f_2 \\ \vdots \\ f_n = \lambda f_{n-1} \\ (\alpha_1 f_1 + \cdots + \alpha_n f_n) + \lambda f_n = -g. \end{cases} \tag{4.25}$$

Let $f_1 = q$, and then it follows from the first $(n-1)$ equations of system (4.25) that $f_2 = \lambda q$, $f_3 = \lambda^2 q, \ldots, f_n = \lambda^{n-1} q$ and from the last equation we obtain a relation

$$-q(\alpha_1 + \lambda\alpha_2 + \cdots + \lambda^{n-1}\alpha_n + \lambda^n) = g.$$

Taking into account that α_1 are coefficients of the characteristic polynomial $\alpha(s)$ of the matrix A, we can write the last relation as

$$-q\alpha(\lambda) = g. \tag{4.26}$$

If $\lambda \in \text{spec}\{A\}$, then $g = 0$ and q can be arbitrary. Now if $\lambda \notin \text{spec}\{A\}$, then, for any g, there exists a unique $q = -g/\alpha(\lambda) \neq 0$ corresponding to equation (4.26).

In this case, the vector F has the form

$$F = q(1, \lambda, \ldots, \lambda^{n-1}).$$

Thus, if the pair $\{C, A\}$ is given in the canonical form, then we can use scalar observers in order to reconstruct the functionals $\sigma = Fx$ generated by the vectors F which are collinear with the vectors $F(\lambda) = (1, \lambda, \lambda^2, \ldots, \lambda^{n-1})$, where $\lambda < 0$. Note that if $\lambda \in \text{spec}\{A\}$, then $F(\lambda)$ is an eigenvector of the matrix A corresponding to λ.

Let $\lambda_1, \lambda_2, \ldots, \lambda_{n-1} \in \mathbb{R}$, $\lambda_i \neq \lambda_j$ for $i \neq j$, $\lambda_i < 0$ $(i = 1, \ldots, n-1)$. Then, every λ can be associated with the vector $F(\lambda_i)$ and every functional $\sigma_i = F(\lambda_i)x$ can be reconstructed by a scalar observer. In addition, the vectors $F(\lambda_1)$, $F(\lambda_2), \ldots,$

$F(\lambda_{n-1})$ and the vector $C = (0, \dots, 0, 1)$ form a basis in the space $\mathbb{R}^n$. This follows from the nondegeneracy of the matrix

$$\begin{pmatrix} 1 & \lambda_1 & \lambda_1^2 & \dots & \lambda_1^{n-2} & \lambda_1^{n-1} \\ 1 & \lambda_2 & \lambda_2^2 & \dots & \lambda_2^{n-2} & \lambda_2^{n-1} \\ \dots & \dots & \dots & \dots & \dots & \dots \\ 1 & \lambda_{n-1} & \lambda_{n-1}^2 & \dots & \lambda_{n-1}^{n-2} & \lambda_{n-1}^{n-1} \\ 0 & 0 & 0 & \dots & 0 & 1 \end{pmatrix} = \begin{pmatrix} F(\lambda_1) \\ F(\lambda_2) \\ \dots \\ F(\lambda_{n-1}) \\ C \end{pmatrix}$$

for which the principal minor of order $(n-1)$ is a Vandermonde determinant and is nondegenerate if $\lambda_i \neq \lambda_j$ $(i \neq j)$.

If we are given an arbitrary vector $F \in \mathbb{R}^{1\times n}$, then this vector can be expanded, uniquely, according to the indicated basis

$$F = \sum_{i=1}^{n-1} \eta_i F(\lambda_i) + \eta_n C. \tag{4.27}$$

In this case, the functional $\sigma = Fx$ has the form

$$\sigma(t) = \sum_{i=1}^{n-1} \eta_i \sigma_i(t) + \eta_n y(t).$$

Since every functional $\sigma_i(t)$ $(i = 1, \dots, n-1)$ can be reconstructed by a scalar observer and $y(t)$ is a known output, it follows that the following statement holds true.

Theorem 4.13. *Suppose that system* (4.1) *is observable and the pair* $\{C, A\}$ *is given in a canonical form. Suppose, furthermore, that we are given a set of real numbers* λ_i, $i = 1, \dots, n-1$, *such that* $\lambda_i < 0$, $\lambda_i \neq \lambda_j$ *for* $i \neq j$. *Then, for every* λ_i *the vector* $F(\lambda_i) = (1, \lambda_i, \dots, \lambda_i^{n-1})$ *is defined and the vectors* $\{F(\lambda_1), \dots, F(\lambda_{n-1}), C\}$ *form a basis in* $\mathbb{R}^n$.

The functional $\sigma = Fx$, *where* $F \in \mathbb{R}^{1\times n}$, *can be reconstructed by an observer of order* k, *where* k *is the number of nonzero coefficients* η_i $(i = 1, \dots, n-1)$ *in expansion* (4.27) *of the vector* F *according to the indicated basis. In this case* $k \leq n-1$.

Remark 4.14. If, in the conditions of the theorem, we take a set of complex numbers, i.e., $\lambda_i \in \mathbb{C}$, $\mathrm{Re}(\lambda_i) < 0$ $(i = 1, \dots, n-1)$, then, as before, every λ_i will be associated with the vector $F(\lambda_i) \in \mathbb{C}^{1\times n}$. The vectors $\{F(\lambda_1), \dots, F(\lambda_{n-1}), C\}$ form a basis in $\mathbb{C}^n$ if λ_i are different. Moreover, it is easy to see that the complex-conjugate values λ and $\bar{\lambda}$ are associated with the complex-conjugate vectors, i.e.,

$$\overline{F(\lambda)} = F(\bar{\lambda}).$$

Let us consider now, in greater detail, the possibility of lowering the order of the observer for the given vector F. For this purpose we must find a set λ for which the number of nonzero coefficients η_i $(i = 1, \dots, n-1)$ in expansion (4.27) is minimal.

Suppose that we are given a vector $F = (f_1, \dots, f_n)$. It follows from Theorem 4.13 that the functional $\sigma = Fx$ can be reconstructed by an observer of order $k < n-1$ if there exists a set $\{\lambda_1, \dots, \lambda_k\}$, $\lambda_i < 0$, $\lambda_i \neq \lambda_j$ such that the vector F is decomposed according to the vectors $F(\lambda_1), \dots, F(\lambda_k)$ and C (i.e., in expansion (4.27) only the first k from $(n-1)$ coefficients η_i are nonzero). This condition is fulfilled if

$$\operatorname{rank}\begin{pmatrix} F \\ F(\lambda_1) \\ \vdots \\ F(\lambda_k) \\ C \end{pmatrix} = k + 1. \tag{4.28}$$

Taking into account the explicit expressions for the vectors F, $F(\lambda_i)$, and C, we write condition (4.28) as

$$\operatorname{rank}\begin{pmatrix} f_1 & f_2 & \dots & f_{n-1} & f_n \\ 1 & \lambda_1 & \dots & \lambda_1^{n-2} & \lambda_1^{n-1} \\ \dots & \dots & \dots & \dots & \dots \\ 1 & \lambda_k & \dots & \lambda_k^{n-2} & \lambda_k^{n-1} \\ 0 & 0 & \dots & 0 & 1 \end{pmatrix} = k + 1.$$

This condition is equivalent to the condition

$$\operatorname{rank}\begin{pmatrix} f_1 & f_2 & \dots & f_{n-1} \\ 1 & \lambda_1 & \dots & \lambda_1^{n-2} \\ \dots & \dots & \dots & \dots \\ 1 & \lambda_k & \dots & \lambda_k^{n-2} \end{pmatrix} = \operatorname{rank} H(F, \lambda_1, \dots, \lambda_k) = k$$

obtained from the preceding condition by removing the last row and last column from the matrix being investigated.

Now we carry out the transformations of the matrix $H(F, \lambda_1, \dots, \lambda_k)$ which do not change its rank: from every column, beginning with the second, we subtract the preceding column multiplied by λ_i. As a result we obtain a matrix

$$\begin{pmatrix} f_1 & f_2 - \lambda_1 f_1 & f_3 - \lambda_1 f_2 & \dots & f_{n-1} - \lambda_1 f_{n-2} \\ 1 & 0 & 0 & \dots & 0 \\ 1 & \lambda_2 - \lambda_1 & (\lambda_2 - \lambda_1)\lambda_2 & \dots & (\lambda_2 - \lambda_1)\lambda_2^{n-3} \\ \dots & \dots & \dots & \dots & \dots \\ 1 & (\lambda_k - \lambda_1) & (\lambda_k - \lambda_1)\lambda_k & \dots & (\lambda_k - \lambda_1)\lambda_k^{n-3} \end{pmatrix}.$$

We subtract the second row from the following one, and then, taking into account that $\lambda_i \neq \lambda_j$, divide the corresponding rows by $(\lambda_i - \lambda_1) \neq 0$. As a result, we obtain a matrix

$$\begin{pmatrix} f_1 & f_2 - \lambda_1 f_1 & f_3 - \lambda_1 f_2 & \dots & f_{n-1} - \lambda_1 f_{n-2} \\ 1 & 0 & 0 & \dots & 0 \\ 0 & 1 & \lambda_2 & \dots & \lambda_2^{n-3} \\ \hdotsfor{5} \\ 0 & 1 & \lambda_k & \dots & \lambda_k^{n-3} \end{pmatrix}.$$

From the first row we subtract the second row multiplied by f_1 and then interchange the places of the first two rows. As a result of all these transformations which do not change the rank of the matrix being investigated, we obtain a matrix

$$\left(\begin{array}{c|cccc} 1 & 0 & 0 & \dots & 0 \\ \hline 0 & f_2 - \lambda_1 f_1 & f_3 - \lambda_1 f_2 & \dots & f_{n-1} - \lambda_1 f_{n-2} \\ 0 & 1 & \lambda_2 & \dots & \lambda_2^{n-3} \\ \hdotsfor{5} \\ 0 & 1 & \lambda_k & \dots & \lambda_k^{n-3} \end{array}\right). \tag{4.29}$$

The isolated submatrix which is in the rows $(2, 3, \dots, k)$ and columns $(2, 3, \dots, n-1)$ has the same structure as the original matrix $H(F, \lambda_1, \dots, \lambda_k)$, and, consequently, we can carry out for it the same transformations.

For our convenience, we introduce the following notation. We define the functions $P_i(x_1, \dots, x_{i+1})$ recurrently:

$$\begin{aligned} &P_0(x_1) = x_1 \\ &P_{i+1}(x_1, \dots, x_{i+2}) = P_i(x_2, \dots, x_{i+2}) - \lambda_i P_i(x_1, \dots, x_{i+1}). \end{aligned} \tag{4.30}$$

Then matrix (4.29), in the new notation, assumes the form

$$\left(\begin{array}{c|cccc} 1 & 0 & 0 & \dots & 0 \\ \hline 0 & P_1(f_1, f_2) & P_1(f_2, f_3) & \dots & P_1(f_{n-2}, f_{n-1}) \\ 0 & 1 & \lambda_2 & \dots & \lambda_2^{n-3} \\ \hdotsfor{5} \\ 0 & 1 & \lambda_k & \dots & \lambda_k^{n-3} \end{array}\right).$$

Carrying out the transformations described above for the isolated submatrix, we obtain a matrix

$$\begin{pmatrix} 1 & 0 & 0 & 0 & \dots & 0 \\ 0 & 1 & 0 & 0 & \dots & 0 \\ 0 & 0 & P_2(f_1, f_2, f_3) & P_2(f_2, f_3, f_4) & \dots & P_2(f_{n-3}, f_{n-2}, f_{n-1}) \\ 0 & 0 & 1 & \lambda_3 & \dots & \lambda_3^{n-4} \\ \dots & \dots & \dots & \dots & \dots & \dots \\ 0 & 0 & 1 & \lambda_k & \dots & \lambda_k^{n-4} \end{pmatrix}.$$

Performing the indicated procedure k times, we pass to the matrix

$$\begin{pmatrix} 1 & 0 & \dots & 0 & 0 & \dots & 0 \\ 0 & 1 & \dots & 0 & 0 & \dots & 0 \\ \dots & \dots & \dots & \dots & \dots & \dots & \dots \\ 0 & 0 & \dots & 1 & 0 & \dots & 0 \\ 0 & 0 & \dots & 0 & P_k(f_1, \dots, f_{k+1}) & \dots & P_k(f_{n-k-1}, \dots, f_{n-1}) \end{pmatrix}.$$

It is obvious that the indicated matrix of order $(k+1) \times (n-1)$ has rank k if and only if the following relations hold:

$$\begin{cases} P_k(f_1, \dots, f_{k+1}) = 0 \\ P_k(f_2, \dots, f_{k+2}) = 0 \\ \vdots \\ P_k(f_{n-k-1}, \dots, f_{n-1}) = 0. \end{cases} \tag{4.31}$$

Let us investigate in greater detail the multivariable function $P_k(x_1, \dots, x_{k+1})$ and consider a polynomial of degree k

$$\varphi_k(s) = \prod_{i=1}^{k}(s - \lambda_i) = s^k + l_k s^{k-1} + \dots + l_1. \tag{4.32}$$

We have the following lemma.

Lemma 4.15. *The function of* $(k+1)$ *variables* $P_k(x_1, \dots, x_{k+1})$ *defined recurrently by eqs.* (4.30) *has the form*

$$P_k(x_1, \dots, x_{k+1}) = l_1 x_1 + l_2 x_2 + \dots + l_k x_k + x_{k+1}, \tag{4.33}$$

where l_i *are coefficients of the polynomial* $\varphi_k(s)$ *from* (4.32).

Proof. Let us prove the statement of the lemma by induction. For $k = 1$ we are given one value λ_1, with

$$\varphi_1(s) = s - \lambda_1 = l_1 + s,$$

i.e., $l_1 = -\lambda_1$. It follows from (4.30) that

$$P_0(x_1) = x_1$$
$$P_1(x_1, x_2) = P_0(x_2) - \lambda_1 P_0(x_1) = x_2 - \lambda_1 x_1 = l_1 x_1 + x_2,$$

i.e., the statement is valid for $k = 1$.

Suppose that it is valid for $(k-1)$. We shall show that it is also valid for k. Consider the polynomial $\varphi_k(s)$

$$\begin{aligned}\varphi_k(s) = \varphi_{k-1}(s)(s-\lambda_k) &= (s^{k-1} + \tilde{l}_{k-1}s^{k-2} + \cdots + \tilde{l}_1)(s-\lambda_k)\\ &= s^k + s^{k-1}(\tilde{l}_{k-1} - \lambda_k) + s^{k-2}(\tilde{l}_{k-2} - \lambda_k \tilde{l}_{k-1}) + \cdots\\ &\quad \cdots + s(\tilde{l}_1 - \lambda_k \tilde{l}_2) + (-\tilde{l}_1\lambda_k),\end{aligned} \tag{4.34}$$

where $\tilde{l}_i$ denote the corresponding coefficients of the polynomial $\varphi_{k-1}(s) = \prod_{i=1}^{k-1}(s-\lambda_i)$. It follows from (4.34) that

$$l_i = \tilde{l}_i - \lambda_k \tilde{l}_{i+1}, \quad i = 1, \dots, k \quad (\tilde{l}_k = 1,\ \tilde{l}_0 = 0). \tag{4.35}$$

Since relation (4.33) is valid for $(k-1)$, it follows from the recurrent relations (4.30) and (4.35) that

$$\begin{aligned}&P_k(x_1, \dots, x_{k+1}) = P_{k-1}(x_2, \dots, x_{k+1}) - \lambda_k P_{k-1}(x_1, \dots, x_k)\\ &= (\tilde{l}_1 x_2 + \tilde{l}_2 x_3 + \cdots + \tilde{l}_{k-1}x_k + x_{k+1}) - \lambda_k(\tilde{l}_1 x_1 + \tilde{l}_2 x_2 + \cdots + \tilde{l}_{k-1}x_{k-1} + x_k)\\ &= (-\lambda_k \tilde{l}_1)x_1 + (\tilde{l}_1 - \lambda_k \tilde{l}_2)x_2 + \cdots + (\tilde{l}_{k-2} - \lambda_k \tilde{l}_{k-1})x_{k-2} + (\tilde{l}_{k-1} - \lambda_k)x_{k-1} + x_{k+1}\\ &= l_1 x_1 + l_2 x_2 + \cdots + l_{k-2}x_{k-2} + l_{k-1}x_k + x_{k+1},\end{aligned}$$

i.e., the relation is satisfied for k. The lemma is proved. □

Taking into account now the explicit expression for the function $P_k(x_1, \dots, x_{k+1})$, we can write equations (4.31) in the form

$$\begin{pmatrix} l_1 & l_2 & \dots & l_k & 1 & 0 & \dots & 0\\ 0 & l_1 & \dots & l_{k-1} & l_k & 1 & \dots & 0\\ \vdots & \vdots & \ddots & & & \ddots & \ddots & \vdots\\ 0 & 0 & \dots & l_1 & & \dots & l_k & 1 \end{pmatrix} \begin{pmatrix} f_1\\ f_2\\ \vdots\\ f_{n-1}\end{pmatrix} = 0,$$

where l_i are coefficients of the polynomial of order k with roots $\lambda_1, \dots, \lambda_k$. We can regard the latter equations as equations for l_i, namely,

$$\begin{pmatrix} f_1 & f_2 & \dots & f_k\\ f_2 & f_3 & \dots & f_{k+1}\\ \dots & \dots & \dots & \dots\\ f_{n-k-1} & f_{n-k} & \dots & f_{n-2}\end{pmatrix} \begin{pmatrix} l_1\\ l_2\\ \vdots\\ l_k\end{pmatrix} = -\begin{pmatrix} f_{k+1}\\ f_{k+2}\\ \vdots\\ f_{n-1}\end{pmatrix}. \tag{4.36}$$

From our reasoning and Theorem 4.13 we have the following theorem.

Theorem 4.16. *Let system* (4.1) *be given in a canonical form,* $l = 1$. *The functional* $\sigma = Fx$, *where* $F = (f_1, \dots, f_n) \in \mathbb{R}^{1\times n}$, *can be reconstructed by an observer of order k if, among the solutions of system* (4.36), *there exists a column* $L = (l_1, \dots, l_k)^\top$ *corresponding to the Hurwitz polynomial* $\varphi_k(s) = l_1 + l_2 s + \dots + l_k s^{k-1} + s^k$ *which has different real roots.*

In order to construct an observer on the basis of Theorem 4.16, we have to find the indicated solution of system (4.36), column L, and also the values λ_i, $i = 1, \dots, k$, corresponding to it. We put the vector $F(\lambda_i)$ and functional $\sigma_i = F(\lambda_i)x$ into correspondence with every λ_i. All functionals σ_i can be reconstructed by scalar observers and the functional $\sigma = Fx$ can be decomposed with respect to the functionals $\sigma_1, \dots, \sigma_k$ and y.

Note that equations (4.36) coincide completely with equations (4.18).

Thus, Theorem 4.16 is a special case of Theorem 4.2. The conditions imposed on the Hurwitz column L in Theorem 4.16 (namely, the requirement that the polynomial corresponding to the vector L should have different real roots) is connected with the method of proving Theorem 4.16 and may be removed (in this case Theorems 4.2 and 4.16 will give the same sufficient condition for the existence of an observer which is also a necessary condition).

In order to remove the requirement of different real roots of the polynomial $\varphi_k(s) = l_1 + l_2 s + \dots + l_k s^{k-1} + s^k$, we shall consider the case where this polynomial has multiple roots or complex-conjugate roots. Let us show that, in this case as well, with the aid of small changes of the algorithm of synthesizing the observer, on the basis of scalar observers, we can solve the given problem.

4.4.1 The case of multiple roots

We begin with the case where the polynomial $\varphi_k(s)$ has a root λ of multiplicity m. We shall prove an auxiliary statement.

Lemma 4.17. *Suppose that the pair* $\{C, A\}$ *is observable, the vector* $F(\lambda)$ *satisfies the equation*

$$F(\lambda)(A - \lambda I) = gC$$

for a given λ *and a certain* g. *Also suppose that the feedback vector* $D \in \mathbb{R}^{n\times 1}$ *is such that* $\lambda \in \operatorname{spec}\{A - DC\}$. *Then* $F(\lambda)$ *is the left eigenvector of the matrix* $(A - DC) = A_D$ *corresponding to the eigenvalue* λ.

Proof. Let λ be defined. We assume for simplifying the proof that the pair $\{C, A\}$ is in an observable canonical form. Then, as was shown above, the vector $F(\lambda)$ has the form

$$F(\lambda) = q(1, \lambda, \dots, \lambda^{n-1})^\top.$$

Suppose that the vector $D = (d_1, \dots, d_n)^\top$ is such that $\lambda \in \operatorname{spec}\{A - DC\}$. Then it follows from the explicit representation (4.7) for A and G that

$$\det(sI - A_D) = s^n + (\alpha_n + d_n)s^{n-1} + \dots + (\alpha_1 + d_1).$$

Since $\lambda \in \operatorname{spec}\{A_D\}$, it follows that

$$\lambda^n + (\alpha_n + d_n)\lambda^{n-1} + \dots + (\alpha_1 + d_1) = 0$$

and then

$$F(\lambda)(A - DC) = q \begin{pmatrix} 1 \\ \lambda \\ \lambda^2 \\ \vdots \\ \lambda^{n-1} \end{pmatrix}^\top \begin{pmatrix} 0 & 0 & \dots & 0 & -(\alpha_1 + d_1) \\ 1 & 0 & \dots & 0 & -(\alpha_2 + d_2) \\ 0 & 1 & \dots & 0 & -(\alpha_3 + d_3) \\ \cdots & \cdots & \cdots & \cdots & \cdots \\ 0 & 0 & \dots & 1 & -(\alpha_n + d_n) \end{pmatrix}$$

$$= q \begin{pmatrix} \lambda \\ \lambda^2 \\ \vdots \\ \lambda^{n-1} \\ -(\alpha_1 + d_1) - (\alpha_2 + d_2)\lambda - \dots - (\alpha_n + d_n)\lambda^{n-1} \end{pmatrix}^\top$$

$$= q \begin{pmatrix} \lambda \\ \lambda^2 \\ \vdots \\ \lambda^n \end{pmatrix}^\top = \lambda F(\lambda).$$

The lemma is proved. □

Since the pair $\{C, A\}$ is observable, the spectrum of the matrix $A_D = A - DC$ is fully defined by the choice of the vector D. We choose the vector D such that

$$\det(sI - A_D) = \varphi_k(s)\tilde{\varphi}_{n-k}(s),$$

where $\varphi_k(s)$ is a Hurwitz polynomial defined above and $\tilde{\varphi}_{n-k}(s)$ is a polynomial of order $(n - k)$ which has no roots in common with $\varphi_k(s)$.

Let λ be a root of multiplicity m of the polynomial $\varphi_k(s)$ (and of the polynomial $\det(sI - A_D)$, respectively). In this case, the matrix A_D can be reduced to the Jordan form by a nondegenerate transformation of coordinates, and, since the pair $\{C, A\}$ is observable, the real eigenvalue λ will be associated with exactly one Jordan cell of size m.

We denote by $F_1(\lambda)$ the eigenvector of the matrix A_D corresponding to λ. In the original basis (corresponding to the canonical form of observability) we have

$F_1(\lambda) = (1, \lambda, \dots, \lambda^{n-1})$. We denote by $F_2(\lambda), \dots, F_m(\lambda)$ the root vectors of the matrix A_D which correspond to λ. These vectors are defined by the relations

$$F_{i+1}(\lambda)A_D = \lambda F_{i+1}(\lambda) + F_i(\lambda), \quad i = 1, 2, \dots, m-1.$$

Note that if $\lambda_1, \dots, \lambda_p$ are real roots of the polynomial $\varphi_k(s)$ of multiplicity $m_1, \dots, m_p$, respectively, $m_1 + \dots + m_p = k$, then the vectors $F_1(\lambda_1), \dots, F_{m_1}(\lambda_1)$, $F_1(\lambda_2), \dots, F_{m_p}(\lambda_p)$ are linearly independent since they form a part of the Jordan basis for the matrix A_D.

In addition to the vectors $F_i(\lambda)$, where $\lambda < 0$, we shall consider the functionals $\sigma_i(\lambda) = F_i(\lambda)x$. Since $F_1(\lambda)$ is an eigenvector of the matrix $A_D = A - DC$, the functional $\sigma_1(\lambda)$ satisfies the equation

$$\begin{aligned}\dot{\sigma}_1 &= F_1(\lambda)\dot{x} = F_1(\lambda)Ax = F_1(\lambda)(A - DC)x + F_1(\lambda)DCx \\ &= \lambda\sigma_1 + (F_1(\lambda)D)y,\end{aligned}$$

and, consequently, if $\lambda < 0$, then the functional $\sigma_1(\lambda)$ can be reconstructed by the scalar observer

$$\dot{\tilde{\sigma}}_1 = \lambda\tilde{\sigma}_1 + (F_1(\lambda)D)y. \tag{4.37}$$

The functional $\sigma_2(\lambda) = F_2(\lambda)x$ satisfies the equation

$$\dot{\sigma}_2 = F_2(\lambda)Ax = F_2(\lambda)(A - DC)x + F_2(\lambda)DCx = \lambda\sigma_2 + \sigma_1 + (F_2(\lambda)D)y.$$

Since (4.37) gives an exponentially converging estimate $\tilde{\sigma}_1$ of the functional σ_1, it follows that we can use the observer

$$\dot{\tilde{\sigma}}_2 = \lambda\tilde{\sigma}_2 + \tilde{\sigma}_1 + (F_2(\lambda)D)y \tag{4.38}$$

for reconstructing $\sigma_2(\lambda)$. Thus, two scalar observers (4.37) and (4.38) together give an estimate for two functionals σ_1 and σ_2.

Continuing the indicated procedure, we can construct estimates for the other functionals $\sigma_i(\lambda)$. Let the estimate $\tilde{\sigma}_i(\lambda)$ be known. Since

$$\dot{\sigma}_{i+1} = \lambda\sigma_{i+1} + \sigma_i + (F_{i+1}(\lambda)D)y,$$

we can use the observer

$$\dot{\tilde{\sigma}}_{i+1} = \lambda\tilde{\sigma}_{i+1} + \tilde{\sigma}_i + (F_{i+1}(\lambda)D)y \tag{4.39}$$

for reconstructing the functional $\sigma_{i+1}(\lambda)$.

It follows from the arguments given above that if $\varphi_k(s)$ has real roots $\lambda_1, \dots, \lambda_p$ of multiplicities $m_1, \dots, m_p$ respectively, then the system of observers of form (4.37)–(4.39) reconstructs exponentially the functionals

$$\sigma_1(\lambda_1), \ \dots, \ \sigma_{m_1}(\lambda_1), \ \sigma_1(\lambda_2), \ \dots, \ \sigma_{m_p}(\lambda_p), \tag{4.40}$$

and the number of the employed scalar observers coincides with the number of functionals and is equal to k.

Let us return to the problem of reconstructing the arbitrary defined functional $\sigma = Fx$. Since the system of functionals (4.40) is reconstructed by an observer of order k (i.e., k scalar observers), and the output $y = Cx$ is known, it follows that for reconstructing the functional $\sigma = Fx$ by an observer of order k it is sufficient that this functional should be decomposable according to system (4.40) and y, or, otherwise,

$$F = \sum_{i=1}^{p}\sum_{j=1}^{m_p} \gamma_{ij} F_j(\lambda_i) + \gamma_{k+1} y.$$

The last condition holds if

$$\operatorname{rank}\begin{pmatrix} F \\ F_1(\lambda_1) \\ \vdots \\ F_{m_p}(\lambda_p) \\ C \end{pmatrix} = k + 1. \tag{4.41}$$

Let us consider the last condition in greater detail. We introduce the following notation. Let

$$F_i(\lambda) = (f_i^1(\lambda), f_i^2(\lambda), \dots, f_i^n(\lambda)),$$

i.e., $f_i^j(\lambda)$ is the jth coordinate of the vector $F_i(\lambda)$. Then, by virtue of the definition of the vectors $F_i(\lambda)$, as well as the explicit representation of the matrices A and C, we have relations

$$\begin{aligned} f_1^{j+1}(\lambda) - \lambda f_1^j(\lambda) &= 0 \\ f_{i+1}^{j+1}(\lambda) - \lambda f_{i+1}^j(\lambda) &= f_i^j(\lambda) \\ f_1^{j+1}(\lambda) - \lambda' f_1^j(\lambda) &= (\lambda - \lambda') f_1^j(\lambda) \\ f_{i+1}^{j+1}(\lambda) - \lambda' f_{i+1}^j(\lambda) &= (\lambda - \lambda') f_{i+1}^j(\lambda) + f_i^j(\lambda), \quad i = 1, \dots, m(\lambda) - 1, \end{aligned} \tag{4.42}$$

where $m(\lambda)$ is the multiplicity of the root λ. For the matrix from condition (4.41) we carry out transformations similar to those carried out for the case of simple roots. Now we take into account that $C = (0, \dots, 0, 1)$, and therefore we can write condition

(4.41) in the form

$$\operatorname{rank}\begin{pmatrix} f_1 & f_2 & f_3 & \dots & f_{n-1} \\ f_1^1(\lambda_1) & f_1^2(\lambda_1) & f_1^3(\lambda_1) & \dots & f_1^{n-1}(\lambda_1) \\ f_2^1(\lambda_1) & f_2^2(\lambda_1) & f_2^3(\lambda_1) & \dots & f_2^{n-1}(\lambda_1) \\ \dots & \dots & \dots & \dots & \dots \\ f_{m_1}^1(\lambda_1) & f_{m_1}^2(\lambda_1) & f_{m_1}^3(\lambda_1) & \dots & f_{m_1}^{n-1}(\lambda_1) \\ f_1^1(\lambda_2) & f_1^2(\lambda_2) & f_1^3(\lambda_2) & \dots & f_1^{n-1}(\lambda_2) \\ \dots & \dots & \dots & \dots & \dots \\ f_{m_p}^1(\lambda_p) & f_{m_p}^2(\lambda_p) & f_{m_p}^3(\lambda_p) & \dots & f_{m_p}^{n-1}(\lambda_p) \end{pmatrix} = k. \tag{4.43}$$

We take into account the explicit expression for the vectors

$$F_1(\lambda_i) = (1, \lambda_i, \lambda_i^2, \dots, \lambda_i^{n-1}).$$

Then condition (4.43) assumes the form

$$\operatorname{rank}\begin{pmatrix} f_1 & f_2 & f_3 & \dots & f_{n-1} \\ 1 & \lambda_1 & \lambda_1^2 & \dots & \lambda_1^{n-2} \\ f_2^1(\lambda_1) & f_2^2(\lambda_1) & f_2^3(\lambda_1) & \dots & f_2^{n-1}(\lambda_1) \\ \dots & \dots & \dots & \dots & \dots \\ f_{m_1}^1(\lambda_1) & f_{m_1}^2(\lambda_1) & f_{m_1}^3(\lambda_1) & \dots & f_{m_1}^{n-1}(\lambda_1) \\ 1 & \lambda_2 & \lambda_2^2 & \dots & \lambda_2^{n-2} \\ \dots & \dots & \dots & \dots & \dots \\ f_{m_p}^1(\lambda_p) & f_{m_p}^2(\lambda_p) & f_{m_p}^3(\lambda_p) & \dots & f_{m_p}^{n-1}(\lambda_p) \end{pmatrix} = k.$$

Let us carry out transformations which do not change the rank of this matrix by analogy with the case of simple roots. From every column, beginning with the second, we subtract the preceding column multiplied by λ_1. Taking into account relations (4.42), we obtain a matrix

$$\begin{pmatrix} f_1 & f_2 - \lambda_1 f_1 & \dots & f_{n-1} - \lambda_1 f_{n-1} \\ 1 & 0 & \dots & 0 \\ f_2^1(\lambda_1) & 1 & \dots & \lambda_1^{n-3} \\ f_3^1(\lambda_1) & f_2^1(\lambda_1) & \dots & f_2^{n-2}(\lambda_1) \\ \dots & \dots & \dots & \dots \\ f_{m_1}^1(\lambda_1) & f_{m_1-1}^1(\lambda_1) & \dots & f_{m_1-1}^{n-2}(\lambda_1) \\ 1 & (\lambda_2 - \lambda_1) & \dots & (\lambda_2 - \lambda_1)\lambda_2^{n-3} \\ f_2^1(\lambda_2) & (\lambda_2 - \lambda_1) f_2^1(\lambda_2) + 1 & \dots & (\lambda_2 - \lambda_1) f_2^{n-2}(\lambda_2) + \lambda_2^{n-3} \\ \dots & \dots & \dots & \dots \\ f_{m_p}^1(\lambda_p) & \begin{pmatrix} (\lambda_p - \lambda_1) f_{m_p}^1(\lambda_p) \\ + f_{m_p-1}^1(\lambda_p) \end{pmatrix} & \dots & \begin{pmatrix} (\lambda_p - \lambda_1) f_{m_p}^{n-2}(\lambda_p) \\ + f_{m_p-1}^{n-2}(\lambda_p) \end{pmatrix} \end{pmatrix}.$$

Then we subtract the second row, multiplied by the corresponding coefficient, from all the other rows. We divide the rows corresponding to the vectors $F_1(\lambda_i)$ by $(\lambda_i - \lambda_1) \neq$

$0, i > 1$ (the roots λ_i are different). After this procedure we shall successively divide the row corresponding to $F_j(\lambda_i)$, $j \geq 2$, by $(\lambda_i - \lambda_1)$ and subtract it from the following row. If, as before, we denote $P_0(x_1) = x_1$, $P_1(x_1, x_2) = P_0(x_1) - \lambda_1 P_0(x_1) = x_2 - \lambda_1 x_1$ and interchange the first two rows of the transformed matrix, we obtain a matrix of the form

$$\begin{pmatrix} 1 & 0 & 0 & \dots & 0 \\ 0 & P_1(f_1, f_2) & P_1(f_2, f_3) & \dots & P_1(f_{n-2}, f_{n-1}) \\ 0 & 1 & \lambda_1 & \dots & \lambda_1^{n-3} \\ 0 & f_2^1(\lambda_1) & f_2^2(\lambda_1) & \dots & f_2^{n-2}(\lambda_1) \\ \dots & \dots & \dots & \dots & \dots \\ 0 & f_{m_1-1}^1(\lambda_1) & f_{m_1-1}^2(\lambda_1) & \dots & f_{m_1-1}^{n-2}(\lambda_1) \\ 0 & 1 & \lambda_2 & \dots & \lambda_2^{n-3} \\ \dots & \dots & \dots & \dots & \dots \\ 0 & f_{m_p}^1(\lambda_p) & f_{m_p}^2(\lambda_p) & \dots & f_{m_p}^{n-2}(\lambda_p) \end{pmatrix}.$$

The minor of this matrix located in the rows $(2, 3, \dots, k)$ and columns $(2, 3, \dots, n-1)$ has the same structure as the original matrix, and, consequently, we can perform for it the transformations described above. This case differs from the case of simple roots by the fact that the first m_1 steps are performed with the same coefficient λ_1, the following m_2 steps with coefficient λ_2, and so on. After performing $m_1 + m_2 + \dots + m_p = k$ steps, the matrix assumes the form

$$\begin{pmatrix} 1 & 0 & \dots & 0 & 0 & \dots & 0 \\ 0 & 1 & \dots & 0 & 0 & \dots & 0 \\ \dots & \dots & \dots & \dots & \dots & \dots & \dots \\ 0 & 0 & \dots & 1 & 0 & \dots & 0 \\ 0 & 0 & \dots & 0 & P_k(f_1, \dots, f_{k+1}) & \dots & P_k(f_{n-k-1}, \dots, f_{n-1}) \end{pmatrix},$$

where, as before, the function

$$P_k(x_1, \dots, x_{k+1}) = x_{k+1} + l_k x_k + \dots + l_1,$$

where l_i are coefficients of the polynomial

$$\varphi_k(s) = \prod_{j=1}^{p} (s - \lambda_j)^{m_j} = s^k + l_k s^{k-1} + \dots + l_1.$$

Thus, the rank condition takes the standard form

$$\begin{cases} P_k(f_1, \dots, f_{k+1}) = 0 \\ \vdots \\ P_k(f_{n-k-1}, \dots, f_{n-1}) = 0, \end{cases}$$

or, with due account of the explicit representation for $P_k(x_1, \dots, x_k)$, the form of system (4.36) with the only difference that now l_1 are coefficients of a polynomial with multiple roots. Thus, we have the following corollary for Theorem 4.16.

Corollary 4.18. *Theorem* 4.16 *remains valid if the vector* $L' = (l_1, \dots, l_k)^\top$, *which is a solution of system* (4.36), *corresponds to the Hurwitz polynomial with real (possibly multiple) roots.*

4.4.2 The case of complex roots

Theorem 4.16 remains valid in the case where the components of the vector L' correspond to the Hurwitz polynomial with complex-conjugate roots. In this case, we can construct the observer using the method of scalar observers.

For definiteness, we shall consider the case of a simple pair of complex-conjugate roots λ and $\bar{\lambda}$. The case where we have multiple complex-conjugate roots can be considered according to the scheme similar to the case of multiple real roots, but this leads to cumbersome computations and therefore we omit the details.

Let us consider the second-order Hurwitz polynomial $\alpha(s) = s^2 + \alpha_2 s + \alpha_1$ whose roots are λ and $\bar{\lambda}$. It is associated with the companion matrix

$$\Lambda = \begin{pmatrix} 0 & 1 \\ -\alpha_1 & -\alpha_2 \end{pmatrix}.$$

Let the matrix $F_\lambda \in \mathbb{R}^{2\times n}$ satisfy the equation

$$F_\lambda A - \Lambda F_\lambda = KC, \tag{4.44}$$

where $K \in \mathbb{R}^{2\times 1}$. Then the two-dimensional linear functional $\sigma_\lambda = F_\lambda x \in \mathbb{R}^{2\times 1}$ satisfies the equation

$$\dot{\sigma}_\lambda = \Lambda \sigma_\lambda + Ky,$$

and, consequently, since Λ is a Hurwitz matrix, its asymptotic estimate is given by the two-dimensional observer

$$\dot{\tilde{\sigma}}_\lambda = \Lambda \tilde{\sigma}_\lambda + Ky. \tag{4.45}$$

By a nondegenerate transformation with a complex matrix $P \in \mathbb{C}^{2\times 2}$ the matrix Λ can be reduced to the diagonal form

$$P^{-1}\Lambda P = \begin{pmatrix} \lambda & 0 \\ 0 & \bar{\lambda} \end{pmatrix} = \Lambda'.$$

We denote $F' = P^{-1}F_\lambda \in \mathbb{C}^{2\times n}$, $K' = P^{-1}K \in \mathbb{C}^{2\times 1}$. Then, after the transformation with the matrix P^{-1}, equation (4.44) assumes the form

$$F'A - \Lambda' F' = K'C.$$

Since the matrix Λ' is diagonal, we can regard the last equation as a system of two equations relative to the rows F_1' and F_2' of the matrix F', namely,

$$\begin{aligned} F_1'(A - \lambda I) &= K_1' C \\ F_2'(A - \bar{\lambda} I) &= K_2' C, \end{aligned} \tag{4.46}$$

where K_i', $i = 1, 2$, are coefficients of the matrix K'. It follows from (4.46) that with the accuracy to the coefficients

$$F_1' = K_1' F(\lambda), \quad F_2' = K_2' F(\bar{\lambda}),$$

where, as before, $F(\lambda) = (1, \lambda, \dots, \lambda^{n-1})$. Since $F' = P^{-1} F_\lambda$, the complex-conjugate rows $F(\lambda)$ and $F(\bar{\lambda})$ are expressed in terms of the rows of the matrix $F_\lambda \in \mathbb{R}^{2\times n}$.

Thus, if, for the given functional $\sigma = Fx$, system (4.36) has a solution $L' = (l_1, \dots, l_k)^\top$, where l_i are coefficients of the Hurwitz polynomial with a pair of complex-conjugate roots λ and $\bar{\lambda}$, then the vector F can be decomposed with respect to the k vectors which include the complex-conjugate vectors $F(\lambda)$ and $F(\bar{\lambda})$ (this follows from the proof of Theorem 4.2 and Corollary 4.9). However, in this case, there exists a real matrix $F_\lambda \in \mathbb{R}^{2\times n}$ such that the functional $\sigma_\lambda = F_\lambda x$ can be reconstructed by a two-dimensional observer and $F(\lambda)$ and $F(\bar{\lambda})$ are expressed in terms of the rows of the matrix F_λ. Consequently, in the decomposition of F the complex rows $F(\lambda)$ and $F(\bar{\lambda})$ can be replaced by the real rows of the matrix F_λ.

A similar procedure can be carried out for all pairs of complex-conjugate roots of the polynomial $\varphi_k(s)$. Now if the polynomial $\varphi_k(s)$ has multiple conjugate roots λ and $\bar{\lambda}$ of multiplicity m, then the pairs of complex-conjugate root vectors $\{F_i(\lambda)\ F_i(\bar{\lambda})\}$ are also replaced by the corresponding two-dimensional real matrices. We omit the details.

Thus, we have the following corollary.

Corollary 4.19. *Theorem* 4.16 *is valid if the solution of system* (4.36), *which is the vector* L', *is a Hurwitz vector.*

Thus, in the case of a scalar functional and scalar output the methods of pseudoinputs and scalar observers give the same results.

4.5 Systems with vector output

Let us generalize the method of scalar observers to the case of systems with vector output, i.e., consider system (4.1) in the case where $y = Cx \in \mathbb{R}^l$, $l > 1$.

Consider the problem of reconstruction of the scalar functional $\sigma = Fx$, $F \in \mathbb{R}^{1\times n}$. We assume that the pair $\{C, A\}$ is observable, $\operatorname{rank}\binom{F}{C} = l + 1$, and the observability index is equal to ν. Then system (4.1) can be reduced, by means of a nondegenerate transformation of coordinates and outputs, to the canonical Luenberger form

$$\begin{cases} \dot{x}_i = A_{ii}x_i + \sum_{j=1, j\neq i}^{l} \bar{a}_{ij} y_j + B_i u, \quad i = 1,\dots,l; \quad x_i \in \mathbb{R}^{\nu_i} \\ y_i = \bar{C}_i x_i, \end{cases} \tag{4.47}$$

where $\nu = \max \nu_i$, $\nu_1 + \dots + \nu_l = n$; the pairs $\{C_i, A_{ii}\}$ are observable and given in the canonical form

$$A_{ii} = \begin{pmatrix} 0 & 0 & \dots & 0 & * \\ 1 & 0 & \dots & 0 & * \\ \dots & \dots & \dots & \dots & \dots \\ 0 & 0 & \dots & 1 & * \end{pmatrix}, \quad \bar{C}_i = (0,\dots,0,1),$$

where $*$ are, possibly, nonzero elements of the matrix A_{ii}. We denote by

$$\bar{v}_i = \sum_{j=1, j\neq i}^{l} \bar{a}_{ij} y_j + B_i u, \quad i = 1,\dots,l,$$

the known input signals of the subsystems from (4.47). Then (4.47) can be regarded as l independent systems with scalar outputs y_i

$$\dot{x}_i = A_{ii}x_i + \bar{v}_i, \quad y_i = \bar{C}_i x_i, \quad i = 1,\dots,l. \tag{4.48}$$

In the new basis the functional σ has the form

$$\sigma = Fx = \sum_{i=1}^{l} F_i x_i, \quad F_i = (f_1^i, \dots, f_{\nu_i}^i). \tag{4.49}$$

Since the inputs $\bar{v}_i$ are known and their influence in the observer can always be compensated, we can assume, in what follows, without loss of generality, that $\bar{v}_i \equiv 0$, $i = 1,\dots,l$.

By analogy with the scalar case, we shall consider linear functionals which can be reconstructed by scalar observers. They are functionals $\sigma_\lambda = \tilde{F}(\lambda)x$, where the vector $\tilde{F}(\lambda) \in \mathbb{R}^{1\times n}$ satisfies the equation

$$\tilde{F}(\lambda)A = \lambda\tilde{F}(\lambda) + DC$$

where $D \in \mathbb{R}^{1\times l}$ is a constant matrix, $\lambda < 0$. As in the scalar case, by a direct verification we can find an explicit expression for $\tilde{F}(\lambda)$:

$$\tilde{F}(\lambda) = (F_1(\lambda), F_2(\lambda), \dots, F_l(\lambda))$$
$$F_i(\lambda) = \gamma_i(\lambda)(1, \lambda, \lambda^2, \dots, \lambda^{\nu_i - 1}), \quad F_i(\lambda) \in \mathbb{R}^{1\times\nu_i}.$$

Here $\gamma_i(\lambda)$ are constants dependent on the parameters of the system, the chosen λ, and the row D; by the choice of D the value of $\gamma_i(\lambda)$ can be defined arbitrarily; $F_i(\lambda)$ is a vector-row of length ν_i corresponding to the ith scalar subsystem from (4.48).

As in the scalar case, we choose k different real values of the parameter λ (i.e., $\lambda_1, \lambda_2, \dots, \lambda_k$, $\lambda_j < 0$, $\lambda_i \neq \lambda_j$) and try to decompose the vector F from (4.49) by the system of vectors $\tilde{F}(\lambda_j)$ and $C_i = (0, \dots, 0, \bar{C}_i, 0, \dots, 0) \in \mathbb{R}^{1\times n}$ $(i = 1, \dots, l)$. For F to be decomposable according to the indicated system of vectors, it is necessary and sufficient that the following rank condition be fulfilled:

$$\operatorname{rank}\begin{pmatrix} F \\ \tilde{F}(\lambda_1) \\ \tilde{F}(\lambda_2) \\ \vdots \\ \tilde{F}(\lambda_k) \\ C_1 \\ \vdots \\ C_l \end{pmatrix} = k + l.$$

With due account of the explicit representation of the vectors, the last condition has the form

$$\operatorname{rank}\begin{pmatrix} F_1 & F_2 & \dots & F_l \\ \gamma_{11}F_1(\lambda_1) & \gamma_{12}F_2(\lambda_1) & \dots & \gamma_{1l}F_l(\lambda_1) \\ \gamma_{21}F_1(\lambda_2) & \gamma_{22}F_2(\lambda_2) & \dots & \gamma_{2l}F_l(\lambda_2) \\ \dots & \dots & \dots & \dots \\ \gamma_{k1}F_1(\lambda_k) & \gamma_{k2}F_2(\lambda_k) & \dots & \gamma_{kl}F_l(\lambda_k) \\ \bar{C}_1 & 0 & \dots & 0 \\ 0 & \bar{C}_2 & \dots & 0 \\ \dots & \dots & \dots & \dots \\ 0 & 0 & \dots & \bar{C}_l \end{pmatrix} = \operatorname{rank} M = k + l. \tag{4.50}$$

Here $\gamma_{ij} = \gamma_j(\lambda_i)$ are constants defined arbitrarily.

Note that the vectors $F_i(\lambda_j)$, $j = 1, \dots, k$, form the Vandermonde matrix

$$\begin{pmatrix} F_i(\lambda_1) \\ F_i(\lambda_2) \\ \vdots \\ F_i(\lambda_k) \end{pmatrix} = \begin{pmatrix} 1 & \lambda_1 & \dots & \lambda_1^{\nu_i-1} \\ 1 & \lambda_2 & \dots & \lambda_2^{\nu_i-1} \\ \dots & \dots & \dots & \dots \\ 1 & \lambda_k & \dots & \lambda_k^{\nu_i-1} \end{pmatrix},$$

and therefore, if the set $\lambda_1, \dots, \lambda_k$, $\lambda_i \neq \lambda_j$, $k \leq \nu - 1$ is defined, the vectors

$$\tilde{F}(\lambda_j) = (F_1(\lambda_j), F_2(\lambda_j), \dots, F_l(\lambda_j)), \quad j = 1, \dots, k,$$

are linearly independent.

For the rank condition (4.50) to be fulfilled, it is necessary and sufficient that the rank condition

$$\operatorname{rank}\begin{pmatrix} F_i \\ F_i(\lambda_1) \\ \vdots \\ F_i(\lambda_k) \\ \bar{C}_i \end{pmatrix} = \operatorname{rank} M_i = k+1, \quad i = 1,\dots,l, \tag{4.51}$$

should be fulfilled for each subsystem. Indeed, if conditions (4.51) are fulfilled, then F_i is expressed in terms of $F_i(\lambda_i),\dots,F_i(\lambda_k)$ and $\bar{C}_i$, and, consequently, by the choice of γ_{ij} we can achieve a situation when condition (4.50) is fulfilled.

Note that conditions (4.51) will be fulfilled for all subsystems if $k = \nu - 1$, where the observability index $\nu = \max \nu_i$ is the dimension of the maximal subsystem. Thus, for a system with vector output we can construct an observer of order $(\nu - 1)$ with any predefined real spectrum.

Taking into account the explicit representations in the canonical basis for $F_i(\lambda_j)$ and $\bar{C}_i$, we can write conditions (4.51) in the form

$$\operatorname{rank} M_i = \operatorname{rank}\begin{pmatrix} f_1^i & f_2^i & \dots & f_{\nu_i}^i \\ 1 & \lambda_1 & \dots & \lambda_1^{\nu_i-1} \\ \dots & \dots & \dots & \dots \\ 1 & \lambda_k & \dots & \lambda_k^{\nu_i-1} \\ 0 & 0 & \dots & 1 \end{pmatrix} = k+1, \quad i = 1,\dots,l,$$

or, what is the same,

$$\operatorname{rank} M_i' = \begin{pmatrix} f_1^i & f_2^i & \dots & f_{\nu_i-1}^i \\ 1 & \lambda_1 & \dots & \lambda_1^{\nu_i-2} \\ \dots & \dots & \dots & \dots \\ 1 & \lambda_k & \dots & \lambda_k^{\nu_i-2} \end{pmatrix} = k, \quad i = 1,\dots,l.$$

Carrying out for the matrices M_i' the transformations described in detail for the scalar case, we obtain equations for the components of the vector F

$$\begin{pmatrix} f_1^i & f_2^i & \dots & f_k^i \\ f_2^i & f_3^i & \dots & f_{k+1}^i \\ \dots & \dots & \dots & \dots \\ f_{\nu_i-k-1}^i & f_{\nu_i-k}^i & \dots & f_{\nu_i-2}^i \end{pmatrix}\begin{pmatrix} l_1 \\ l_2 \\ \vdots \\ l_k \end{pmatrix} = -\begin{pmatrix} f_{k+1}^i \\ f_{k+2}^i \\ \vdots \\ f_{\nu_i-1}^i \end{pmatrix}, \quad i = 1,\dots,l. \tag{4.52}$$

Note that the vector $L' = (l_1,\dots,l_k)^\top$ is the general solution of all l systems from (4.52) since this vector defines the polynomial $\varphi_k(s) = \prod_{j=1}^k (s-\lambda_j) = s^k + l_k s^{k-1} + \dots + l_1$. Thus, we have the following theorem.

Theorem 4.20. *Suppose that system* (4.1) *is observable,* $l > 1$, *the pair* $\{C, A\}$ *is given in the canonical form* (2.11). *The functional* $\sigma = Fx$, *where* $F \in \mathbb{R}^{1\times n}$, $\operatorname{rank}\binom{F}{C} = l + 1$,

$$F = (F_1, \dots, F_l), \quad F_i = (f_1^i, \dots, f_{\nu_i}^i),$$

can be reconstructed by an observer of order k *if, among the solutions of the system of linear equations*

$$\left(\begin{array}{cccc} f_1^1 & f_2^1 & \cdots & f_k^1 \\ f_2^1 & f_3^1 & \cdots & f_{k+1}^1 \\ \cdots & \cdots & \cdots & \cdots \\ f_{\nu_1-k-1}^1 & f_{\nu_1-k}^1 & \cdots & f_{\nu_1-2}^1 \\ \hline f_1^2 & f_2^2 & \cdots & f_2^k \\ \cdots & \cdots & \cdots & \cdots \\ f_{\nu_2-k-1}^2 & f_{\nu_2-k}^2 & \cdots & f_{\nu_2-2}^k \\ \hline \cdots & \cdots & \cdots & \cdots \\ f_{\nu_l-k-1}^l & f_{\nu_l-k}^l & \cdots & f_{\nu_l-2}^l \end{array}\right) \begin{pmatrix} l_1 \\ l_2 \\ \vdots \\ l_k \end{pmatrix} = - \left(\begin{array}{c} f_{k+1}^1 \\ f_{k+2}^1 \\ \vdots \\ f_{\nu_1-1}^1 \\ \hline f_{k+1}^2 \\ \vdots \\ f_{\nu_2-1}^2 \\ \hline \vdots \\ f_{\nu_l-1}^l \end{array}\right), \tag{4.53}$$

there exists a Hurwitz column $L' = (l_1, \dots, l_k)^\top$ *corresponding to a polynomial with real and distinctive roots.*

Remark 4.21. As in the case of a scalar output, the requirement of reality of the roots of the polynomial $\varphi_k(s) = s^k + l_k s^{k-1} + \dots + l_1$ as well as the requirement of the absence of multiple roots of this polynomial can be removed. To do this, as in the scalar case, we have to consider in greater detail the case of multiple roots and the case of complex-conjugate roots according to the scheme described above. We omit the details.

We shall show now that the conditions of Theorem 4.20 are not only sufficient but also necessary for constructing a k-order functional observer.

Suppose that an observer of this kind exists. Then it is described by a system of linear differential equations

$$\begin{cases} \dot{z} = Pz + Qu + Ry \\ \tilde{\sigma} = Tz + Gy, \end{cases} \tag{4.54}$$

where $z \in \mathbb{R}^k$, $P \in \mathbb{R}^{k\times k}$, $Q \in \mathbb{R}^{k\times m}$, $R \in \mathbb{R}^{k\times l}$, $T \in \mathbb{R}^{1\times k}$, and $G \in \mathbb{R}^{l\times l}$ are constant matrices. For simplicity, we shall consider the case where the eigenvalues of the matrix P are real and different. Moreover, for observer (4.54) to give an estimate of the functional $\sigma(t)$, the matrix P must be a Hurwitz matrix. Thus we have

$$\operatorname{spec}\{P\} = \{\lambda_1, \dots, \lambda_k\}; \quad \lambda_i < 0, \quad i = 1, \dots, k; \quad \lambda_i \neq \lambda_j.$$

In this case, by a nondegenerate transformation, observer (4.54) can be reduced to the form where the matrix P is diagonal. Therefore, without loss of generality, we assume, in what follows, that $P = \operatorname{diag}(\lambda_1, \dots, \lambda_k)$.

The author of [87] gives conditions which should be imposed on the matrices P, Q, R, T, and G for which observer (4.54) gives an exponential estimate of the functional $\sigma(t)$

$$\begin{aligned} &F = TH + GC, \\ &Q = HB, \\ &HA - PH = RC, \\ &P \text{ is a Hurwitz matrix,} \end{aligned} \tag{4.55}$$

where $H \in \mathbb{R}^{k\times n}$ is a constant matrix such that z is an asymptotic estimate of the functional Hx.

In the case of a diagonal matrix P for $\lambda_i < 0$ the last condition from (4.55) is fulfilled. The second condition from (4.55) is fulfilled by the requisite choice of the matrix Q. Suppose that $H_1, \dots, H_k$ are rows of the matrix H and $t_1, \dots, t_k$ are components of the row T. Then the first condition from (4.55) means that

$$F = \sum_{i=1}^{k} t_i H_i + GC,$$

i.e., the row F is linearly expressed in terms of the rows $H_1, \dots, H_k$ and the rows of the matrix C, and, consequently, the condition

$$\operatorname{rank}\begin{pmatrix} F \\ H_1 \\ \vdots \\ H_k \\ C \end{pmatrix} = k + l \tag{4.56}$$

is fulfilled. In addition, since the structure of the matrix P is diagonal, the third condition from (4.55) can be written as a system of equations for $H_1, \dots, H_k$, namely,

$$\begin{cases} H_1(A - \lambda_1 I) = R_1 C \\ H_2(A - \lambda_2 I) = R_2 C \\ \qquad\vdots \\ H_k(A - \lambda_k I) = R_k C, \end{cases}$$

where $R_1, \dots, R_k$ are rows of the matrix R. It follows from (4.56) that, with an accuracy to within the transformations, we have

$$H_i = \tilde{F}(\lambda_i),$$

and, consequently, conditions (4.56) have the form

$$\operatorname{rank}\begin{pmatrix} F \\ \tilde{F}(\lambda_1) \\ \vdots \\ \tilde{F}(\lambda_k) \\ C \end{pmatrix} = k + l.$$

In this way, the condition is fulfilled, which, after the transformations which do not change the rank of the matrix, gives condition (4.53) of Theorem 4.20.

We can present the same arguments for the case where the spectrum of the matrix P has a more complicated structure. We omit the details.

Thus, we have the following theorem.

Theorem 4.22. *Suppose that system* (4.1) *is observable for* $l > 1$*, the pair* $\{C, A\}$ *is in the canonical Luenberger form. The functional* $\sigma = Fx$*, where* $F \in \mathbb{R}^{1\times n}$, $\operatorname{rank}\binom{F}{C} = l + 1$,

$$F = (F_1, \dots, F_l), \quad F_i = (f_1^i, \dots, f_{\nu_i}^i),$$

can be reconstructed by an observer of order k if and only if among the solutions of system (4.53) *there exists a Hurwitz column* $L' = (l_1, \dots, l_k)^\top$.

4.6 Analysis of properties of solutions of linear systems of special type

We can see from the theorems given in this chapter that in order to solve the problem of the existence of observers of order k, we have to study the properties of solutions of a certain system of linear equations ((4.18), (4.22) or (4.53) depending on the dimensions of the functional and the output of the original system). To be more precise, we have to find out whether the given system of linear equations is solvable and whether there are Hurwitz solutions of this system. In this case, properly, we consider not one equation but a family of equations for different k. In this subsection we shall analyze the properties of families of this kind for a scalar functional and a scalar output.

Let us consider in greater detail a family of systems of algebraic equations of form (4.18) for different k and prove a number of auxiliary statements which connect the existence or absence of solutions (Hurwitz solutions) of the system of equations (4.18) for different k.

The first question that arises in the analysis of system (4.18) is as follows: if there exists a solution of a system of form (4.18) for some k^*, then whether there exists a solution for $k > k^*$. The following lemma answers this question.

Lemma 4.23. *If the system of algebraic equations* (4.18) *is solvable for a certain k, then the system is also solvable for* $(k+1)$.

Proof. Suppose that for k system (4.18) has the vector $l = (l_1, \dots, l_k)^\top$ as its solution, i.e.,

$$\underbrace{\begin{pmatrix} f_1 & f_2 & \cdots & f_k \\ f_2 & f_3 & \cdots & f_{k+1} \\ \vdots & \vdots & \vdots & \vdots \\ f_{n-k-1} & f_{n-k} & \cdots & f_{n-2} \end{pmatrix}}_{H_k} \underbrace{\begin{pmatrix} l_1 \\ l_2 \\ \vdots \\ l_k \end{pmatrix}}_{l} = - \underbrace{\begin{pmatrix} f_{k+1} \\ f_{k+2} \\ \vdots \\ f_{n-1} \end{pmatrix}}_{h_k}. \tag{4.57}$$

Consider a column

$$l' = \begin{pmatrix} 0 \\ l \end{pmatrix} \in \mathbb{R}^{k+1}.$$

We shall show that it is a solution of system (4.18) for $(k+1)$, i.e., show that the relation

$$H_{k+1} l' = -h_{k+1}$$

is satisfied. Let us write this system in a more extended form:

$$\begin{pmatrix} f_1 & f_2 & \cdots & f_{k+1} \\ f_2 & f_3 & \cdots & f_{k+2} \\ \vdots & \vdots & \vdots & \vdots \\ f_{n-k-2} & f_{n-k-1} & \cdots & f_{n-2} \end{pmatrix} \begin{pmatrix} 0 \\ l_1 \\ \vdots \\ l_k \end{pmatrix} = - \begin{pmatrix} f_{k+2} \\ f_{k+3} \\ \vdots \\ f_{n-1} \end{pmatrix}.$$

Taking into account the explicit representation of the vector l', we can rewrite the last system in the form

$$\begin{pmatrix} f_2 & \cdots & f_{k+1} \\ f_3 & \cdots & f_{k+2} \\ \vdots & \vdots & \vdots \\ f_{n-k-1} & \cdots & f_{n-2} \end{pmatrix} \begin{pmatrix} l_1 \\ l_2 \\ \vdots \\ l_k \end{pmatrix} = - \begin{pmatrix} f_{k+2} \\ f_{k+3} \\ \vdots \\ f_{n-1} \end{pmatrix}. \tag{4.58}$$

However, system (4.58) is a shortened system (4.57) (without the first equation) and, consequently, (4.58) is obviously compatible if l is a solution of (4.57). The lemma is proved. □

Lemma 4.23 implies a simple corollary.

Corollary 4.24. *If for the family of systems of algebraic equations* (4.18) *there exists a number* k^* *such that for all* $k < k^*$ *the system does not have any solutions and for* k^* *system* (4.18) *has a solution, then for all* $k \geq k^*$ *system* (4.18) *has a solution.*

Note that k^* is the minimal number for which the rank condition (4.20) is fulfilled. The procedure of finding k^* can be easily algorithmized.

The next question that arises when we study the family of systems of algebraic equations (4.18) is the question of existence of the number k^*, i.e., whether at least one of systems of form (4.18) is solvable. Naturally we consider systems for $k = 1, 2, \ldots, n-2$. The answer to this question is negative.

Example 4.25. Consider the vector $F = (0, 0, \ldots, 0, 1, *)$. As can be seen from system (4.18), H_k and h_k do not depend on the last coordinate of the vector F, i.e., on f_n, and therefore of interest is the shortened vector $F' = (f_1, \ldots, f_{n-1}) \in \mathbb{R}^{1\times(n-1)}$. In this example $F' = (0, \ldots, 0, 1)$, f_n is an arbitrary number.

In this case, for all $k = 1, \ldots, n-2$, the matrix H_k and column h_k have the form

$$H_k = \begin{pmatrix} 0 & \ldots & 0 \\ 0 & \ldots & 0 \\ \vdots & \vdots & \vdots \\ 0 & \ldots & 0 \end{pmatrix} \in \mathbb{R}^{(n-k-1)\times k}, \quad h_k = \begin{pmatrix} 0 \\ \vdots \\ 0 \\ 1 \end{pmatrix} \in \mathbb{R}^{k\times 1}.$$

Obviously, the system $H_k l = -h_k$ is not compatible for any k, i.e., there does not exist k^* for the indicated vector F.

The next question that we should consider is the following: if there exists k^* for family (4.18) is this solution of the system unique for k^* and is the solution of the system unique for $k > k^*$?

The following lemma gives the answer to the first part of the question.

Lemma 4.26. *Suppose that $k^* > 0$ is a number such that a system of form* (4.18) *does not have a solution for any $k < k^*$ and has a solution for $k = k^*$. Then, if $k^* \leq \frac{n-1}{2}$, then this solution is unique. Now if $k^* > \frac{n-1}{2}$, then there are infinitely many solutions for k^*.*

Proof. Let $k^* \leq \frac{n-1}{2}$. This means that for k^* the number of equations $(n - k^* - 1)$ for system (4.18) is not less than the number of unknown k^*. Let us assume that for k^* there exist infinitely many solutions of equation (4.18). We denote rank $H_{k^*} = r_{k^*}$ and then

$$\operatorname{rank} H_{k^*} = \operatorname{rank}(H_{k^*} \mid h_{k^*}) = r_{k^*} < k^*.$$

This follows from the explicit form of system (4.18) for k^*:

$$\begin{pmatrix} f_1 & \ldots & f_{k^*} \\ f_2 & \ldots & f_{k^*+1} \\ \vdots & \vdots & \vdots \\ f_{n-k^*-1} & \ldots & f_{n-1} \end{pmatrix} \begin{pmatrix} l_1 \\ l_2 \\ \vdots \\ l_{k^*} \end{pmatrix} = - \begin{pmatrix} f_{k^*+1} \\ f_{k^*+2} \\ \vdots \\ f_{n-1} \end{pmatrix}.$$

Since the systems of equations (4.18) are incompatible for all $k < k^*$, the inequality

$$\operatorname{rank} H_k = r_k < \operatorname{rank}(H_k \mid h_k)$$

is satisfied for $k < k^*$. Let us consider the structure of the matrices H_k, H_{k+1} and $H_k' = (H_k \mid h_k)$ in greater detail:

$$H_k = \begin{pmatrix} f_1 & \cdots & f_k \\ \vdots & \ddots & \vdots \\ f_{n-k-1} & \cdots & f_{n-2} \end{pmatrix} \in R^{(n-k-1)\times k},$$

$$H_k' = \begin{pmatrix} f_1 & \cdots & f_k & f_{k+1} \\ \vdots & \ddots & \ddots & \vdots \\ f_{n-k-2} & \cdots & f_{n-3} & f_{n-2} \\ f_{n-k-1} & \cdots & f_{n-2} & f_{n-1} \end{pmatrix} \in R^{(n-k-1)\times(k+1)},$$

$$H_{k+1} = \begin{pmatrix} f_1 & \cdots & f_k & f_{k+1} \\ \vdots & \ddots & \ddots & \vdots \\ f_{n-k-2} & \cdots & f_{n-3} & f_{n-2} \end{pmatrix} \in R^{(n-k-2)\times(k+1)}.$$

Thus we have

$$\begin{gathered} H_k' = (H_k \mid h_k) = \begin{pmatrix} H_{k+1} \\ h_{k+1}' \end{pmatrix}, \\ h_{k+1}' = (f_{n-k-1}, \ldots, f_{n-2}, f_{n-1}) \in \mathbb{R}^{1\times(k+1)}. \end{gathered} \tag{4.59}$$

Consequently, the matrix H_{k+1} results from the matrix H_k upon the addition of column h_k and removal of the last row of the obtained matrix. Since system (4.18) is compatible for k^*, the column h_{k^*} can be linearly expressed via columns of the matrix H_{k^*}, i.e., there exists a vector l such that

$$H_{k^*} l = -h_{k^*}.$$

Let us take an arbitrary number k such that $0 < k < k^*$ and consider the matrix consisting of the last $(k+1)$ rows of the matrix H_{k^*}:

$$H_{k^*}^{(k+1)} = \begin{pmatrix} f_{n-k^*-k-1} & \cdots & f_{n-k-3} & f_{n-k-2} \\ \vdots & \ddots & \ddots & \vdots \\ f_{n-k^*-1} & \cdots & f_{n-3} & f_{n-2} \end{pmatrix} \in R^{(k+1)\times(k^*)}.$$

Note that $k + 1 \le k^* \le n - k^* - 1$, i.e., the number of rows in the matrix H_{k^*} is not less than $k + 1$ for any $k = 1, \ldots, k^* - 1$, i.e., we can really isolate the matrix $H_{k^*}^{(k+1)}$ from it.

We can see from the explicit representation of $H_{k^*}^{(k+1)}$ that the matrix H_{k+1} can be written in block form

$$H_{k+1} = \begin{pmatrix} (H''_{k+1})^\top \\ (H_{k^*}^{(k+1)})^\top \end{pmatrix},$$

where H''_{k+1} is a matrix of the corresponding dimension. Moreover,

$$h'_{k+1} = (f_{n-k-1} \; \dots \; f_{n-2} \; f_{n-1}) = h^\top_{n-k-2};$$

$$h_{k^*} = \begin{pmatrix} f_{k^*+1} \\ \dots \\ f_{n-k-2} \\ --- \\ f_{n-k-1} \\ \dots \\ f_{n-1} \end{pmatrix} \begin{matrix} \left.\vphantom{\begin{matrix} a \\ a \\ a \end{matrix}}\right\} \mathbb{R}^{(n-k^*-1)-(k+1)} \\ \\ \left.\vphantom{\begin{matrix} a \\ a \\ a \end{matrix}}\right\} \mathbb{R}^{k+1}, \end{matrix}$$

i.e., $(h'_{k+1})^\top$ is the last $(k+1)$-dimensional part of the vector h_{k^*}. Then, taking into account (4.18) for k^*, we have

$$(0 \mid -l^\top) H_{k+1} = (0 \mid -l^\top) \begin{pmatrix} (H''_{k+1})^\top \\ (H_{k^*}^{(k+1)})^T \end{pmatrix} = (-H_{k^*}^{(k+1)} \; l)^\top = h'_{k+1}.$$

Consequently, the row h'_{k+1} can be linearly expressed in terms of the rows H_{k+1} and it follows from (4.50) that

$$\operatorname{rank} H'_k = \operatorname{rank} H_{k+1}, \quad k = 1, \dots, k^* - 1.$$

Then

$$\operatorname{rank} H_k = r_k < \operatorname{rank} H'_k = \operatorname{rank} H_{k+1} = r_{k+1}. \tag{4.60}$$

Thus, if $\operatorname{rank} H_{k^*} = r_{k^*} < k^*$, then it follows from (4.60) that $r_k < k$ for all $k < k^*$. However, in that case $\operatorname{rank} H_1 < 1$ as well, i.e., $\operatorname{rank} H_1 = 0$, and this means that $f_1 = f_2 = \dots = f_{n-2} = 0$, i.e., the vector F has the structure from Example 4.25, but in that case there does not exist a number k^* for which system (4.18) is compatible. We have obtained a contradiction, and, consequently, $r_{k^*} = k^*$ and the system of equations (4.18) for $k^* \le \frac{n-1}{2}$ has a unique solution.

Now if $k^* > n - k^* - 1$, i.e., $k^* > \frac{n-1}{2}$ (the number of unknowns exceeds the number of equations), then

$$\operatorname{rank} H_{k^*} \le \min(k^*, n - k^* - 1) = n - k^* - 1 < k^*$$

and system (4.18) has infinitely many solutions. The lemma is proved. □

Remark 4.27. The number k^* may be arbitrary, from 1 to $n-2$.

Here are the corresponding examples.

Example 4.28. Consider a vector $F = (1, \lambda, \lambda^2, \dots, \lambda^{n-2}, *)$ (the last term is of no importance). In this case

$$H_1 = \begin{pmatrix} 1 \\ \lambda \\ \vdots \\ \lambda^{n-3} \end{pmatrix}, \quad h_1 = \begin{pmatrix} \lambda \\ \lambda^2 \\ \vdots \\ \lambda^{n-2} \end{pmatrix},$$

and the equation

$$H_1(l_1) = -h_1$$

has a unique solution $l_1 = -\lambda$. Thus, we have $k^* = 1$ in this case. Note that $k^* = 1$ if and only if the vector F has the indicated structure for some λ.

Example 4.29. Consider a vector $F = (0, \dots, 0, 1, 2, *)$. Then we have

$$H_k = \begin{pmatrix} 0 & \dots & 0 & 0 \\ \vdots & \vdots & \vdots & \vdots \\ 0 & \dots & 0 & 0 \\ 0 & \dots & 0 & 1 \end{pmatrix}, \quad h_k = \begin{pmatrix} 0 \\ \vdots \\ 0 \\ 1 \\ 2 \end{pmatrix}$$

for all $k \le n-3$ and the system is incompatible. For $k = n-2$ these matrices assume the form

$$H_{n-2} = (0, \dots, 0, 1) \in \mathbb{R}^{1\times(n-2)}, \quad h_k = (2).$$

Then system (4.18) has the form

$$(0 \dots 0\ 1) \begin{pmatrix} l_1 \\ \vdots \\ l_{n-2} \end{pmatrix} = 2,$$

it is compatible, and has an infinite number of solutions: $l_{n-2} = 2, l_1, \dots, l_{n-3}$ can be arbitrary. In this case $k^* = n-2$.

For k^* system (4.18) can have a unique solution as well as infinitely many solutions. Let us show that for all $k > k^*$ the solution not only exists (as follows from Lemma 4.23) but there are infinitely many solutions.

We have the following lemma.

Lemma 4.30. *Suppose that k^* is a number beginning with which systems of families* (4.18) *have solutions. Then, for all $k > k^*$, systems* (4.18) *have infinitely many solutions.*

Proof. Let $k > k^*$, and then $(k-1) \geq k^*$, i.e., system (4.18) has a solution $l \in \mathbb{R}^{k-1}$ for $(k-1)$. In that case, by virtue of Lemma 4.23, the vector

$$l' = \begin{pmatrix} 0 \\ l \end{pmatrix} \in \mathbb{R}^k$$

is a special solution of the nonhomogeneous system (4.18) for the indicated k. Let us consider a homogeneous system relative to the unknown vector l^0 for the indicated k:

$$H_k l^0 = 0, \quad l^0 \in \mathbb{R}^k.$$

It follows from the explicit representation of the system of equations (4.18) for $(k-1)$ that

$$H_{k-1} l = -h_{k-1} \Rightarrow (H_{k-1} | h_{k-1}) \begin{pmatrix} l \\ 1 \end{pmatrix} = \begin{pmatrix} H_k \\ -- \\ h'_k \end{pmatrix} \begin{pmatrix} l \\ 1 \end{pmatrix} = 0,$$

whence we have

$$H_k \begin{pmatrix} l \\ 1 \end{pmatrix} = 0,$$

i.e., the column $l^0 = \binom{l}{1} \in \mathbb{R}^k$ is a special solution of the homogeneous equation. Then all vectors

$$\tilde{l} = l' + t l^0, \quad t \in \mathbb{R},$$

are solutions of the nonhomogeneous equation for $k > k^*$, and since $l^0 \neq 0$, their number is infinite.

The lemma is proved. □

The lemmas that we have proved allow us to formulate the following theorem.

Theorem 4.31. *Suppose that we are given a vector $F = (f_1, \ldots, f_n)$ which defines a family of linear systems of algebraic equations of form* (4.18) *for $k = 1, \ldots, n-2$. In that case, either all systems are incompatible or there exists a number k^* such that the systems are incompatible for all $k < k^*$ and compatible for all $k \geq k^*$. If $k^* \leq \frac{n-1}{2}$, then for k^* the solution of the system of algebraic equations is unique and if $k^* > \frac{ni1}{2}$, then for k^* there exist infinitely many solutions. For all $k > k^*$, each of the systems of family* (4.18) *has infinitely many solutions.*

When we solve a problem of synthesis of an observer, it is not only important that system (4.18) should have a solution but it is also significant that it should be a Hurwitz solution, i.e., the column $l = (l_1, \ldots, l_k)^\top$ should define the Hurwitz polynomial $\varphi_k(s) = s^k + l_k s^{k-1} + \cdots + l_1$.

Let $k^{**} \geq k^*$ be a number such that for $k^* < k^{**}$ system (4.18) does not have any Hurwitz solutions and for $k = k^*$ there exists a Hurwitz solutions. Note that k^{**} may exist for family (4.18) and may not exist.

Here are the corresponding examples.

Example 4.32. Consider the vector $F = (1, \lambda, \lambda^2, \ldots, \lambda^{n-2}, *)$ from Example 4.28. In this case, as was shown above, $k^* = 1$. For $\lambda < 0$ and $k = 1$ system (4.18) of the form

$$\begin{pmatrix} 1 \\ \lambda \\ \vdots \\ \lambda^{n-3} \end{pmatrix} (l_1) = -\begin{pmatrix} \lambda \\ \lambda^2 \\ \vdots \\ \lambda^{n-2} \end{pmatrix}$$

has a Hurwitz solution $l_1 = -\lambda$ (corresponding to the Hurwitz polynomial $\varphi_1(s) = s + l_1$), the functional can be reconstructed by the first-order observer ($p(s)$ is a characteristic polynomial of this observer). Thus, in the indicated case we have $k^{**} = 1$.

Now if $\lambda \leq 0$, then, for all $k \geq 1$, the matrices H_k and h_k have nonnegative coefficients, and, consequently, not for any $k \geq 1$ there is a solution of the system

$$H_k l = -h_k$$

with a column l with positive elements l_1, i.e., the necessary condition of being a Hurwitz column is not fulfilled for any k. Consequently, k^{**} does not exist for $\lambda \geq 0$ for a family of systems of form (4.18) (although there exists $k^* = 1$).

This example allows us to formulate a simple statement.

Statement 4.33. If $k^* = 1$ for the family of equations (4.18) but for $k = 1$ the solution is not a Hurwitz solution, then, for all k such that $1 \leq k \leq n - 2$ the solution is not a Hurwitz solution either.

The next question that we will consider is similar to the question concerning the properties of k^*. Suppose that for k^{**} systems (4.18) have a Hurwitz solution for the first time. Will the system have Hutwitz solutions for any $k > k^{**}$?

The following lemma gives an answer to this question.

Lemma 4.34. *If a system of equations of form* (4.18) *has a Hurwitz solution for k, then it also has a Hurwitz solution for* $(k + 1)$.

Proof. As was shown in the proof of Lemma 4.23, if the vector column l is a solution of system (4.18) for a given k, then the vector

$$l' = \begin{pmatrix} 0 \\ l \end{pmatrix} \in \mathbb{R}^{k+1}$$

is a solution of the system for $(k + 1)$.

Let l be a Hurwitz solution for a given k and let it be associated with the Hurwitz polynomial

$$\varphi_k(s) = s^k + l_k s^{k-1} + \cdots + l_1.$$

Then the column l' is associated with the polynomial

$$\varphi'(s) = s^{k+1} + l_k s^k + \cdots + l_1 s$$

which is not a Hurwitz polynomial (k of its roots coincide with the roots of $\varphi_k(s)$ and lie in $\mathbb{C}_-$ and one root is zero).

However, as follows from Theorem 4.31, for $(k+1)$ the solution of the system is not unique (this follows from the fact that $k \geq k^{**} \geq k^*$). The set of solutions of system (4.18) for $(k+1)$ are the vectors

$$\tilde{l} = l' + l^0,$$

where l^0 is a solution of the homogeneous equation

$$H_{k+1} l^0 = 0.$$

Let us consider some vector l^0 with the first nonzero coordinate, i.e.,

$$l^0 = (l_0^0, l_1^0, \ldots, l_k^0), \quad l_0^0 > 0.$$

This condition is fulfilled, in particular, by the vector

$$l^0 = \begin{pmatrix} l \\ 1 \end{pmatrix} \in \mathbb{R}^{n+1}.$$

Indeed, as was indicated above, l^0 is a solution of the homogeneous equation, and since l is a Hurwitz polynomial, it follows that $l_1 = l_0^0 > 0$.

Let us consider vectors

$$\tilde{l}(\varepsilon) = l' + \varepsilon l^0$$

for small values of the parameter $\varepsilon > 0$, where l^0 is the indicated fixed solution of the homogeneous system. We shall show that for small $\varepsilon > 0$ the column $\tilde{l}(\varepsilon)$ is a Hurwitz column. For our purpose, we shall write for the vector

$$\tilde{l}(\varepsilon) = \begin{pmatrix} \varepsilon l_0^0 \\ l_1 + \varepsilon l_1^0 \\ \vdots \\ l_k + \varepsilon l_k^0 \end{pmatrix}$$

the Hurwitz matrix

$$G(\varepsilon) = \begin{pmatrix} l_k + \varepsilon l_k^0 & 1 & 0 & 0 & \ldots & 0 \\ l_{k-2} + \varepsilon l_{k-2}^0 & l_{k-1} + \varepsilon l_{k-1}^0 & l_k + \varepsilon l_k^0 & 1 & \ldots & 0 \\ \vdots & \vdots & \vdots & \vdots & \vdots & \vdots \\ 0 & 0 & 0 & 0 & \ldots & \varepsilon l_0^0 \end{pmatrix} \in \mathbb{R}^{(k+1)\times(k+1)} \tag{4.61}$$

in explicit form. Note that for $\varepsilon = 0$ the principal minor of order k of the matrix $G(0)$ is a Hurwitz matrix for the column l (i.e., for the polynomial $\varphi_k(s)$). Since l is a Hurwitz column, the determinants of the first k principal minors of the matrix $G(0)$ are positive and since they depend continuously on ε, they preserve their positiveness for all $\varepsilon \in (0, \varepsilon^*)$ for a certain $\varepsilon^* > 0$.

Let us consider the last principal minor of order $(k+1)$ which is a determinant of the matrix $G(\varepsilon)$. It is easy to see that

$$\det G(\varepsilon) = \varepsilon l_0^0 \det(G_k(\varepsilon)),$$

where $G_k(\varepsilon)$ is a principal minor of order k of the matrix $G(\varepsilon)$. However, since

$$\det G_k(\varepsilon) > 0 \quad \text{for} \quad \varepsilon \in (0, \varepsilon^*)$$

and $l_0^0 > 0$, we have

$$\det G(\varepsilon) > 0$$

for all $\varepsilon \in (0, \varepsilon^*)$. Consequently, the Hurwitz conditions are fulfilled for all sufficiently small $\varepsilon > 0$ for the columns $\tilde{l}(\varepsilon)$, and, hence, system (4.18) has a Hurwitz solution for $(k+1)$. The lemma is proved. □

Lemma 4.34 allows us to justify the following statement.

Theorem 4.35. *Suppose that we are given a fully of equations* (4.18) *and that* k^{**} *is a number such that for all* $k < k^{**}$ *system* (4.18) *does not have any Hurwitz solutions and has a Hurwitz solution for* k^{**}. *Then, for all* $k \geq k^{**}$, *a system of form* (4.18) *also has Hurwitz solutions.*

Corollary 4.36. *If, for some number* $k = k'$, *the system of equations* (4.18) *does not have Hurwitz solutions, then it does not have Hurwitz solutions for all* $k < k'$ *either.*

Thus, two numbers k^* and k^{**} are defined for the family of systems (4.18). Beginning with k^* solutions appear for the system and beginning with k^{**} they become Hurwitz solutions.

As was shown in Example 4.32, if $k^* = 1$, then either $k^{**} = 1$ or k^{**} does not exist. For $k^* > 1$ the connection between these two indices is more complicated. Consider the corresponding examples.

Example 4.37. Let us consider a vector $F = (-2, 1, 3, 2, 5)$. In this case, $n = 5$ ($f_5 = 5$ is not included into system (4.18)). Let us find k^*. It is easy to see that there are no solutions for $k = 1$. For $k = 2$ the system of equations (4.18) has the form

$$H_2 l = \begin{pmatrix} -2 & 1 \\ 1 & 3 \end{pmatrix} \begin{pmatrix} l_1 \\ l_2 \end{pmatrix} = -\begin{pmatrix} 3 \\ 2 \end{pmatrix} = -h_2.$$

This system has a unique solution $l = \binom{1}{-1}$ in which case the column l is not a Hurwitz column. For $k = 3$ the system under consideration assumes the form

$$H_3 l = (-2\ 1\ 3)\begin{pmatrix} l_1 \\ l_2 \\ l_3 \end{pmatrix} = -2 = -h_3.$$

This system has infinitely many solutions, in particular, $l = (7, 6, 2)^\top$, which is a Hurwitz solution (it corresponds to the Hurwitz polynomial $\varphi_3(s) = s^3 + 2s^2 + 6s + 7$). So, we have

$$k^* = 2 < k^{**} = 3$$

in this example.

Consequently, in the general case, k^* and k^{**} may not coincide and the connection between them is complicated. However, in some cases we can establish certain connections between k^* and k^{**}. One of them is given by Statement 4.33. Here is one more relationship.

Statement 4.38. Suppose that the number $k^* \le \frac{n-2}{2}$ and that system (4.18) does not have Hurwitz solutions for k^*. Then there are no Hurwitz solutions for $k^* + 1$ either.

Proof. Since $k^* \le \frac{n-2}{2}$, it is obvious that $k^* \le \frac{n-1}{2}$, and, consequently, by virtue of Lemma 4.26, the solution of the system for k^* is unique. Suppose that this solution is the column $l = (l_1, \ldots, l_{k^*})^\top$. Consider the system for $k^* + 1$. It follows from the proof of Lemma 4.26 that

$$\operatorname{rank} H_{k^*} = k^*,$$

and since $k^* \le \frac{n-2}{2}$, we have $n - (k^* + 1) - 1 \ge k^*$ whence it follows that

$$\operatorname{rank} H_{k^*+1} = k^*, \quad H_{k^*+1} \in \mathbb{R}^{(n-(k^*+1)-1)\times(k^*+1)}$$

as well.

Consequently, the homogeneous equation

$$H_{k^*+1}\bar{l} = 0, \quad \bar{l} \in \mathbb{R}^{k^*+1} \tag{4.62}$$

has a one-dimensional space of solutions, and since the vector $l^0 = \binom{l}{1}$ is one of the solutions of the homogeneous equation (4.62), the general solution of the nonhomogeneous equation for $(k^* + 1)$ has the form

$$\tilde{l} = \begin{pmatrix} 0 \\ l \end{pmatrix} + \gamma \begin{pmatrix} l \\ 1 \end{pmatrix}, \quad \gamma \in \mathbb{R}. \tag{4.63}$$

The vector column l is associated with the unstable polynomial $\varphi_k(s) = s^k + l_k s^{k-1} + \cdots + l_1$. Let us consider the polynomial $\tilde{\varphi}(s)$ corresponding to the vector $\tilde{l}$ from (4.63):

$$\tilde{\varphi}(s) = \underbrace{s^{k+1} + \left[l_k s^k + l_{k-1} s^{k-1} + \cdots + l_1 s + 0\right]}_{s\varphi_k(s)} + \underbrace{\gamma\left[s^{k+1} + l_k s^{k-1} + \cdots + l_1\right]}_{\gamma\varphi_k(s)} = \varphi(s)(s+\gamma).$$

It follows that if the polynomial $\varphi_k(s)$ is not a Hurwitz polynomial, then the polynomial $\tilde{\varphi}(s)$ is not a Hurwitz polynomial for any $\gamma \in \mathbb{R}$.

Statement 4.38 is proved. □

Remark 4.39. In Example 4.37 $k^* = 2$ but $k^{**} = 3 = k^* + 1$ since the condition $k^* \le \frac{n-2}{2}$ is not fulfilled in the example ($k^* = 2 > \frac{n-2}{2} = 1.5$ in Example 4.37). Therefore, when we pass from k^* to $k^* + 1$ the rank of the matrix H_{k^*+1} diminishes as compared to the rank of H_{k^*}, the space of solutions of the homogeneous system for $k^* + 1$ is two-dimensional and the arguments from the proof of Statement 4.38 are not suitable here.

It follows from Statement 4.38 that if $k^* \le \frac{n-2}{2}$, i.e., the system of equations (4.18) becomes compatible for this k^* when the number of equations (4.18) $n - k^* - 1$ becomes not smaller than the number of unknowns, then the solution for k^* is unique and if $(k^* + 1)$ does not lead us out of the situation of the overdetermined system (when the number of unknowns does not yet exceed the number of equations), then the presence of a Hurwitz solution is wholly determined by the Hurwitz nature of the initial solution for k^*. The following theorem generalizes this statement.

Theorem 4.40. *Suppose that the number $k^* \le \frac{n-1}{2}$. Then for k^* the system of equations* (4.18) *has a unique solution $l \in \mathbb{R}^{k^* \times 1}$. If l is not a Hurwitz vector, then system* (4.18) *does not have Hurwitz solutions for any $k \in [k^*, n - k^* - 1]$.*

Proof. As in Statement 4.38, we can see that the matrix $H_{k^*} \in \mathbb{R}^{(n-k^*-1)\times k^*}$ is of full rank, i.e.,

$$\operatorname{rank} H_{k^*} = k^*.$$

It is taken into account here that the condition $k^* \le \frac{n-1}{2}$ is equivalent to the condition $(n - k^* - 1) \ge k^*$. It follows from Theorem 4.31 that the solution for k^* is unique.

Consider the set of matrices $H_{k^*}, H_{k^*+1}, \ldots, H_{k^*+q}$ until the condition

$$n - (k^* + i) - 1 \ge k^*, \quad i = 0, 1, \ldots, q, \tag{4.64}$$

is fulfilled. Under this condition, in all matrices from the indicated set the number of rows is not smaller than k^*, and since the matrix H_{k+1} results from H_k upon the "addition" of the column h_k on the right-hand side (which can be expressed linearly via the rows of H_k, i.e., this operation does not change the rank of the matrix) and the "removal" of the row below (this does not change the rank of the matrix either since it does not attract the first k^* basic rows), the conditions

$$\operatorname{rank} H_{k^*} = \operatorname{rank} H_{k^*+1} = \cdots = \operatorname{rank} H_{k^*+q} = k^* \tag{4.65}$$

will be fulfilled, where q, in accordance with (4.64), is defined by the relation

$$q = n - 2k^* - 1. \tag{4.66}$$

The last matrix from the set is of the maximal index

$$k^* + q = n - k^* - 1 = k_{\max}.$$

Let us consider the variety of solutions of each system of equations (4.18) for $k = k^* + i$, $i = 1, \ldots, q$. For each k the solution of this system can be represented as the sum of a particular solution of the nonhomogeneous equation (4.18) and the general solution of the homogeneous system. However, since condition (4.65) is fulfilled for every $k = k^* + i$, the homogeneous system for $k = k^* + i$ has exactly i linearly independent solutions. Let us generalize the arguments from the proof of Statement 4.38 and indicate the set of i linearly independent solutions of this kind:

$$l^1 = \begin{pmatrix} l \\ 1 \\ 0 \\ 0 \\ \vdots \\ 0 \end{pmatrix}, \quad l^2 = \begin{pmatrix} 0 \\ l \\ 1 \\ 0 \\ \vdots \\ 0 \end{pmatrix}, \quad l^3 = \begin{pmatrix} 0 \\ 0 \\ l \\ 1 \\ \vdots \\ 0 \end{pmatrix}, \quad \ldots, \quad l^i = \begin{pmatrix} 0 \\ \vdots \\ 0 \\ l \\ 1 \end{pmatrix}, \tag{4.67}$$
$$l^1, \ldots, l^i \in \mathbb{R}^{k^*+i},$$

where the vector $l = (l_1, \ldots, l_{k^*})^\top$ is the unique solution of the original equation (4.18) for k^*.

As the particular solution of the nonhomogeneous equation for $k = (k^* + i)$ we shall take the vector

$$l' = \begin{pmatrix} 0 \\ \vdots \\ 0 \\ l \end{pmatrix}. \tag{4.68}$$

Then the general solution of the nonhomogeneous equation (4.18) for $(k^* + i)$ will have the form

$$\tilde{l} = l' + \sum_{j=1}^{i} \gamma_j l^j, \quad \gamma_j \in \mathbb{R}.$$

Let the column l be associated with the polynomial $\varphi_k(s) = s^{k^*} + l_{k^*} s^{k^*-1} + \cdots + l_1$. Then, reasoning as we did when proving Statement 4.38, we can show that the column $\tilde{l} \in \mathbb{R}^{k^*+i}$ is associated with the polynomial

$$\tilde{\varphi}(s) = \varphi_k(s)(s^i + \gamma_i s^{i-1} + \cdots + \gamma_1).$$

It follows that if $\varphi_k(s)$ is not a Hurwitz polynomial, then the polynomial $\tilde{\varphi}(s)$ is not a Hurwitz polynomial for any $\gamma_1, \ldots, \gamma_i$ either. This statement is valid for all $k = k^*$, $k^* + 1, \ldots, n - k^* - 1$. The theorem is proved. □

Remark 4.41. When the condition $(n - k^* + i) \geq k^*$ is violated, the rank of the matrix H_{k^*+i} becomes lower than k^* and additional solutions of the homogeneous equation appear in addition to the set (4.67), and, in this case, the theorem is no longer valid (precisely this fact is illustrated by Example 4.37).

Remark 4.42. Statement 4.33 is a special case of Theorem 4.40 for $k^* = 1$. Theorem 4.40 allows us to simplify the analysis of problem concerning the synthesis of the functional observer in the case where $k^* \leq \frac{n-1}{2}$ (i.e., where system (4.18) becomes compatible for the first time under the condition that the number of unknowns is not larger than that of the equations). In this case we have to investigate the Hurwitz nature of the unique solution for k^*. If it is not a Hurwitz solution, then we should continue the investigations beginning with the systems of equations of order $(n - k^*)$.

4.7 Minimal functional observers with a defined spectrum

As we did earlier, we shall consider, for simplicity, the fully determined dynamical system

$$\begin{cases} \dot{x} = Ax \\ y = Cx \end{cases} \tag{4.69}$$

without the known input signal (which can always be compensated in the observer). We shall assume that the pair $\{C, A\}$ is observable and given in the Luenberger canonical form (2.33). Then the system is decomposed into l subsystems of order ν_i. We assume, in addition, that ν_i, which are *Kronecker indices*, are arranged with respect to their increase. We have to construct an asymptotic estimate for the functional of the phase vector

$$\sigma = Fx,$$

with $y \in \mathbb{R}^l$, $\sigma \in \mathbb{R}^p$ and $x \in \mathbb{R}^n$. The Luenberger observer which reconstructs the phase vector of system (4.69) is of order $(n - l)$, and therefore of interest is the construction of an observer of an order lower than $(n - l)$.

When synthesizing functional observers, we can distinguish the following two problems:

• the synthesis of an asymptotic observer with some spectrum but of the minimal possible order,

• the synthesis of an observer of a minimal order with any preassigned spectrum (with any preassigned rate of convergence).

In the preceding sections we considered the first one of these problems. Let us now consider the second one. This statement has already been considered in a number of papers. In particular, it was shown in [105] that the guaranteed order of the observer

$$k(p) = \sum_{i=1}^{\min(p,l)} (\nu_i - 1), \tag{4.70}$$

where ν_i $(i = 1, \dots, l)$ are Kronecker indices of system (4.69) arranged according to their nonincrease. Since $\nu_i \geq \nu_{i+1}$, where $\nu_1 = \nu$ is the observability index of the pair $\{C, A\}$, it follows that for $p = 1$ estimate (4.70) gives $k = \nu - 1$ (for the special case $p = 1$ this result was obtained in [87]) and for $p \geq l$ this estimate gives $k = n - l$, i.e., the order of the Luenberger observer for a full-phase vector.

However, for almost all dynamical systems under consideration this estimate can be perfected.

In order to solve our problem, we shall again use the method of scalar observers described above. Suppose that the system is reduced to the Luenberger canonical form (2.33). Since the connection between the subsystems in this form is realized in terms of the measured outputs of the system, which can always be compensated in the observers, instead of (2.33) we can consider a system without connections, i.e.,

$$\begin{cases} \dot{x}^i = A_{ii}x^i, & i = 1, \dots, l, \quad x^i \in \mathbb{R}^{\nu_i} \\ y_i = \bar{C}_i x^i, & y_i \in \mathbb{R} \end{cases} \tag{4.71}$$

where each pair $\{\bar{C}_i, A_{ii}\}$ is in the canonical form.

In the case of system (4.71) the scalar observer can reconstruct the functional $\sigma = F(\lambda)x$, where the vector $F(\lambda)$ has the form

$$\begin{aligned} F(\lambda) &= (\gamma_1 F_1(\lambda), \gamma_2 F_2(\lambda), \dots, \gamma_l F_l(\lambda)) \\ F_i(\lambda) &= (1, \lambda, \dots, \lambda^{\nu_i - 1}), \quad F_i(\lambda) \in \mathbb{R}^{1\times\nu_i}. \end{aligned} \tag{4.72}$$

Here $F_i(\lambda)$ are vectors corresponding to the ith subsystem from (4.71) and γ_i are arbitrary constants. As for a scalar system, the vectors $F(\lambda)$ are left-hand eigenvectors of the matrix $A_L = A - LC$ which has an eigenvalue λ. Moreover, λ can be an eigenvalue of multiplicity l since it is an eigenvalue of each of the diagonal blocks of the matrix A_L. The matrix $L \in \mathbb{R}^{l\times n}$ has the corresponding block structure

$$L = \begin{pmatrix} L_1 & 0 & \cdots & 0 \\ 0 & L_2 & \cdots & 0 \\ & & \ddots & \\ 0 & 0 & \cdots & L_l \end{pmatrix},$$

where the off-diagonal blocks are zero. The distinctive feature of vector systems is that each λ is associated now with l linearly independent eigenvectors which form the space $\Omega(\lambda)$ (γ_i are, in fact, the coordinates in this space).

Let us consider a set of real numbers λ_i ($i = 1, \dots, \nu - 1$) satisfying the condition

$$\lambda_i < 0, \quad \lambda_i \neq \lambda_j \quad \text{for} \quad i \neq j. \tag{4.73}$$

We choose $L \in \mathbb{R}^{l\times n}$ such that the spectrum of $A_L = A - LC$ contains λ_1, λ_2, $\dots$, $\lambda_{\nu-1}$ and the spectrum of $A_L^i = A_{ii} - L_i\bar{C}_i$ contains λ_1, λ_2, $\dots$, λ_{ν_i-1} (i.e., for the first subsystem of the maximal dimension $\nu_1 = \nu$ we use the whole set λ_1, $\lambda_2, \dots, \lambda_{\nu-1}$ whether for the other subsystems only its part which corresponds to the dimension of the subsystem). In this case, λ_1 is a root of multiplicity l of the matrix A_L and the other λ_j are roots of multiplicity not higher than l. Each λ_j is associated with exactly the number of eigenvectors of the matrix A_L corresponding to its multiplicity, and therefore these vectors correspond to the ith subsystems and have the general form $(0, \dots, 0, F_i(\lambda_j), 0, \dots, 0)$. Together with the vectors $C_i = (0, \dots, 0, \bar{C}_i, 0, \dots, 0)$ they form a basis in the space $\mathbb{R}^n$:

$$\begin{array}{cccc}
(F_1(\lambda_1), 0, \dots, 0) & (0, F_2(\lambda_1), \dots, 0) & \dots & (0, 0, \dots, F_l(\lambda_1)) \\
(F_1(\lambda_2), 0, \dots, 0) & \vdots & \dots & \vdots \\
\vdots & (0, F_2(\lambda_{\nu_1-1}), \dots, 0) & \vdots & \underline{(0, 0, \dots, \bar{C}_l)} \\
(F_1(\lambda_{\nu_1-1}), 0, \dots, 0) & \underline{(0, \bar{C}_2, \dots, 0)} & & \\
\underline{(\bar{C}_1, 0, \dots, 0)} & & &
\end{array} \tag{4.74}$$

Note that in the "first column" in (4.74) are exactly ν_1 vectors (including $(\bar{C}_1, 0, \dots, 0)$) and the other columns contain not more than ν_1 vectors (to be more precise, the "ith column" includes ν_i vectors).

Among the indicated set of vectors there are exactly l vectors corresponding to the eigenvalue λ_1 (the "first row" from (4.74)) they form a subspace $\Omega(\lambda_1) \in \mathbb{R}^n$ of dimension l. The eigenvectors corresponding to $\lambda_2, \dots, \lambda_{\nu-1}$ form subspaces $\Omega(\lambda_2)$, $\dots, \Omega(\lambda_{\nu-1})$ and the vectors $(0, \dots, 0, \bar{C}_i, 0, \dots, 0)$ form the subspace Ω_y corresponding to the measured output $y \in \mathbb{R}^l$. The whole space $\mathbb{R}^n$ is decomposed into a direct sum of subspaces $\Omega(\lambda_1), \Omega(\lambda_2), \dots, \Omega(\lambda_{\nu-1}), \Omega_y$, i.e., any vector $F \in \mathbb{R}^{1\times n}$ is decomposed into a sum

$$F = F(\lambda_1) + \dots + F(\lambda_{\nu-1}) + \tilde{C},$$

where $F(\lambda_i) \in \Omega(\lambda_i)$, $\tilde{C} \in \Omega_y$ (i.e., $\tilde{C} = QC, \tilde{C} \in \mathbb{R}^{1\times n}$, $Q \in \mathbb{R}^{1\times l}$).

Each functional $\sigma_i = F(\lambda_i)x$ is reconstructed by a scalar observer with a scalar spectrum λ_i, with

$$\sigma = Fx = \sum_{i=1}^{\nu-1} \sigma_i + Qy.$$

Thus, we have the following theorem.

Theorem 4.43. *Suppose that the dynamical system* (4.69) *is observable,* $l > 1$. *Then the scalar functional* $\sigma = Fx$ *is reconstructed by an observer with the defined spectrum* (*satisfying condition* (4.73)), *the order of the observer does not exceed* $\nu - 1$, *where* ν *is the observability index of the system.*

Let us now consider the vector functional $\sigma = Fx \in \mathbb{R}^{p\times 1}$, where $F \in \mathbb{R}^{p\times n}$ and $p > 1$. In fact, in this case we have to reconstruct p scalar functionals simultaneously by one observer of order k.

We shall consider the matrix F in the canonical basis in block form

$$F = \begin{pmatrix} F_1^1 & F_2^1 & \dots & F_l^1 \\ F_1^2 & F_2^2 & \dots & F_l^2 \\ \vdots & \vdots & \vdots & \vdots \\ F_1^p & F_2^p & \dots & F_l^p \end{pmatrix} = \begin{pmatrix} F^1 \\ F^2 \\ \vdots \\ F^p \end{pmatrix}, \tag{4.75}$$

where the rows $F^i \in \mathbb{R}^{1\times n}$ are decomposed into subrows $F_j^i \in \mathbb{R}^{1\times \nu_j}$ corresponding to the jth subsystems from the canonical representation (4.71).

In the case of vector functional, we have to construct an observer which simultaneously reconstructs the components $\sigma^i = F^i x$ of this functional. In order to solve the problem, we again use the method of scalar observers.

We shall solve this problem successively for the scalar functionals $\sigma^i = F^i x$ and begin the consideration with $\sigma^1 = F^1 x$. We assume that in the row $F^1 = (F_1^1, F_2^1, \dots, F_l^1)$ the first subrow F_1^1, which corresponds to the first block of the maximal dimension ν_1 from the canonical representation (2.33), is not identically zero. If this is not the fact, then we choose among the rows F^i a row which satisfies this condition and renumber the rows F^i.

Now if all $F_1^i \equiv 0$, then this means that the functional $\sigma = Fx$ does not depend on the first block from (4.71). In this case, the problem reduces to the problem of reconstruction of the functional $\bar{\sigma} = \bar{F}x$, where

$$\bar{F} = \begin{pmatrix} F_2^1 & \cdots & F_l^1 \\ \cdots & \cdots & \cdots \\ F_2^p & \cdots & F_l^p \end{pmatrix} \in \mathbb{R}^{p\times(n-\nu_1)}, \quad F = (0\ ;\ \bar{F})$$

for the reduced system

$$\begin{cases} \dot{x}^i = A_{ii}x^i \\ y_i = \bar{C}_i x^i, \quad i = 2, \dots, l, \end{cases}$$

of order $(n - \nu_1)$ with the $(l - 1)$ outputs. The consideration of this reduced problem can be carried out according to the same scheme

Thus, we assume that $F_1^1 \neq 0$, and moreover suppose that F_1^1 and $\bar{C}_1$ are not collinear (i.e., $F_1^1 \neq \gamma \bar{C}_1$) since otherwise, making a linear transformation $\bar{\sigma}^1 = \sigma^1 - \gamma y_1$ with the known output y_1, we make the corresponding subvector zero for the functional $\bar{\sigma}^1$, and this leads to the case described above. In order to reconstruct the functional $\sigma^1 = F^1 x$, we use the scheme of reconstruction of the scalar functional described above.

For our purpose, we choose a spectrum $\lambda_1, \dots, \lambda_{\nu_1 - 1}$ satisfying conditions (4.73). Then the row F^1 can be represented as the sum

$$F^1 = F(\lambda_1) + F(\lambda_2) + \cdots + F(\lambda_{\nu_1 - 1}) + \tilde{C}, \tag{4.76}$$

where $F(\lambda_i)$ are left-hand eigenvectors of the corresponding matrix A_L, $\tilde{C} = QC$.

Each of the vectors $F(\lambda_i)$ has the following structure:

$$F(\lambda_i) = (\gamma_{1i} F_1(\lambda_i), \gamma_{2i} F_2(\lambda_i), \dots, \gamma_{li} F_l(\lambda_i)),$$
$$F_j(\lambda_i) = (1, \lambda_i, \lambda_i^2, \dots, \lambda_i^{\nu_j - 1}) \in \mathbb{R}^{1 \times \nu_j}.$$

Since we have decomposition (4.76) for the full vector $F^1 \in \mathbb{R}^n$, there exists a decomposition

$$F_1^1 = \gamma_{11} F_1(\lambda_1) + \gamma_{12} F_1(\lambda_2) + \cdots + \gamma_{1(\nu_1 - 1)} F_1(\lambda_{\nu_1 - 1}) + \gamma_{1\nu_1} \bar{C}_1$$

for its subvector F_1^1, or, in more detail,

$$\begin{aligned} F_1^1 &= (f_1^1, \dots, f_{\nu_1}^1) \\ &= \gamma_{11}(1, \lambda_1, \lambda_1, \lambda_1^2, \dots, \lambda_1^{\nu_1 - 1}) + \gamma_{12}(1, \lambda_2, \lambda_2, \lambda_2^2, \dots, \lambda_2^{\nu_1 - 1}) + \dots \\ &\quad + \gamma_{1(\nu_1 - 1)}(1, \lambda_{(\nu_1 - 1)}, \lambda_{(\nu_1 - 1)}, \lambda_{(\nu_1 - 1)}^2, \dots, \lambda_{(\nu_1 - 1)}^{\nu_1 - 1}) + \gamma_{1\nu_1}(0, 0, \dots, 0, 1). \end{aligned} \tag{4.77}$$

Without loss of generality, we can assume that in decomposition (4.77) all coefficients $\gamma_{ij} \neq 0$. For our purpose we shall prove the following auxiliary statement.

Statement 4.44. Suppose that the vectors $F = (f_1, \dots, f_k) \in \mathbb{R}^k$ and $C = (0, \dots, 0, 1) \in \mathbb{R}^k$ are collinear and $F(\lambda) = (1, \lambda, \dots, \lambda^{k-1}) \in \mathbb{R}^k$. Then, for almost all sets $\Lambda = (\lambda_1, \dots, \lambda_{k-1})$: $\lambda_i \neq \lambda_j$ for $i \neq j$ (i.e., except for a manifold of measure zero in the space $\mathbb{R}^{k-1}$ of sets Λ) in decomposition

$$F = \sum_{i=1}^{k-1} \gamma_i F(\lambda_i) + \gamma_k C \tag{4.78}$$

all coefficients γ_i are nonzero.

Proof. First we should note that for the indicated choice of λ_i the vectors $F(\lambda_1), \dots, F(\lambda_{k-1})$ and C form a basis in the space $\mathbb{R}^k$, and therefore the coefficients γ_i in decomposition (4.78) are uniquely defined by the choice of Λ. Let us consider a set Λ such that in decomposition (4.78) there exists at least one zero coefficient. We assume for simplicity that γ_{k-1}. Then

$$\sum_{i=1}^{k-2} \gamma_i F(\lambda_i) + \gamma_k C - F = 0,$$

i.e., the vectors $F(\lambda_1), \dots, F(\lambda_{k-2})$, C and F are linearly dependent. In this case the determinant of the matrix composed of these rows is zero. Let us consider it in more detail:

$$\det \begin{pmatrix} F \\ C \\ F(\lambda_1) \\ \vdots \\ F(\lambda_{k-2}) \end{pmatrix} = \det \begin{pmatrix} f_1 & f_2 & \dots & f_{k-1} & f_k \\ 0 & 0 & \dots & 0 & 1 \\ 1 & \lambda_1 & \dots & \lambda_1^{k-2} & \lambda_1^{k-1} \\ \vdots & & & & \vdots \\ 1 & \lambda_{k-2} & \dots & \lambda_{k-2}^{k-2} & \lambda_{k-2}^{k-1} \end{pmatrix} = p(\lambda_1, \dots, \lambda_{k-2}) = 0,$$

where $p(\lambda_1, \dots, \lambda_{k-2})$ is a polynomial of $(k-2)$ variables $\lambda_1, \dots, \lambda_{k-2}$, and if the vectors F and C are collinear, then $p(\lambda_1, \dots, \lambda_{k-2}) \equiv 0$, otherwise this polynomial is not identically zero, its coefficients are defined by the coordinates f_i of the vector F.

Thus, if in decomposition (4.78) of the vector F with respect to the basis $F(\lambda_1), \dots, F(\lambda_{k-2}), C$ there exists at least one zero coefficient, then some $(k-2)$-dimensional part of the vector Λ is a root of a certain polynomial $p(\cdot)$ (we mean a polynomial of many variables and a "vector" root), the set of these polynomials is finite. It remains to note that the set of these roots forms a set of measure zero.

The statement is proved. □

Remark 4.45. Actually the collection of sets $\Lambda = (\lambda_1, \dots, \lambda_{\nu_1-1})$ such that in the corresponding decomposition (4.77) there exists at least one zero coefficient γ_{1i} $(i = 1, \dots, \nu_1)$ is, in $\mathbb{R}^{\nu_1-1}$, the union of surfaces defined by the equations

$$\begin{cases} p(\lambda_1, \dots, \lambda_{\nu_1-2}) = 0 \\ p(\lambda_1, \dots, \lambda_{\nu_1-3}, \lambda_{\nu_1-1}) = 0 \\ p(\lambda_2, \dots, \lambda_{\nu_1-2}, \lambda_{\nu_1-1}) = 0. \end{cases}$$

It is taken into account here that the polynomial $p(\cdot)$ is "symmetrical" relative to its arguments, i.e., if the set Λ is its root, then any rearrangement in this set is also a root of this polynomial.

It follows from the statement that we have proved that for any set Λ there exists arbitrarily close to it set $\tilde{\Lambda}$ such that all coefficients for it $\gamma_{1i} \neq 0$. In this case we can choose $\tilde{\Lambda}$ such that it should possess the required decree of stability.

Let us return to the decomposition of the full-phase vector F^1 of form (4.76). The first part of the vector F^1, namely, F_1^1, has a decomposition of form (4.77). In this case, by virtue of the inequality to zero of the coefficients γ_{1i}, the first components of the vectors $F(\lambda_i)$, i.e., the subvectors

$$\begin{aligned}
\gamma_{11}(1, \lambda_1, \dots, \lambda_1^{\nu_1-1}) &= \tilde{F}_1(\lambda_1) \\
\gamma_{12}(1, \lambda_2, \dots, \lambda_2^{\nu_1-1}) &= \tilde{F}_1(\lambda_2) \\
&\vdots \\
\gamma_{1(\nu_1-1)}(1, \lambda_{\nu_1-1}, \dots, \lambda_{\nu_1-1}^{\nu_1-1}) &= \tilde{F}_1(\lambda_{\nu_1-1}) \\
\gamma_{1\nu_1}(0, 0, \dots, 0, 1) &= \tilde{\tilde{C}}_1
\end{aligned} \tag{4.79}$$

form a basis in the space $\mathbb{R}^{\nu_1}$ (i.e., in the subspace of the first subsystem from (4.71)).

Each vector $F^2, F^3, \dots, F^p$ of the required functional $\sigma = Fx$ has the form

$$F^i = (F_1^i, \dots, F_l^i),$$

where $F_j^i \in \mathbb{R}^{\nu_j}$ is a part of the vector F^i corresponding to the jth subsystem. Then every one of the first subvectors F_1^i is decomposed uniquely with respect to basis (4.79), i.e.,

$$F_1^i = \sum_{j=1}^{\nu_1-1} \eta_{ij} \tilde{F}_1(\lambda_j) + \eta_{i\nu_1} \tilde{\tilde{C}}_1.$$

Then, taking into account that $\tilde{F}_1(\lambda_i)$ are the first parts of the vectors $F(\lambda_i)$ in decomposition (4.76), we have

$$\tilde{F}^i = F^i - \sum_{j=1}^{\nu_1-1} \eta_{ij} \tilde{F}(\lambda_j) - \eta_{i\nu_1} \tilde{C}_1 = (0, \tilde{F}_2^i, \dots, \tilde{F}_l^i), \tag{4.80}$$

where $\tilde{F}_j^i \in \mathbb{R}^{\nu_j}$ $(j = 2, \dots, l)$ are some vectors, i.e., by a linear transformation with the aid of the vectors $\tilde{F}(\lambda_i)$ (which have been already used for constructing observers for the first component $\sigma^1 = F^1 x$) we can make all first subvectors vanish for all F^i, $i = 2, \dots, p$.

Transformation (4.80) is associated with the transformation of the scalar functionals $\sigma^i = F^i x$, namely,

$$\tilde{\sigma}^i = \tilde{F}^i x = \sigma^i - \sum_{j=1}^{\nu_1-1} \eta_{ij} \sigma(\lambda_j) - \eta_{i\nu_1} \gamma_{1\nu_1} y_1. \tag{4.81}$$

In this case, each functional $\sigma(\lambda_j)$ is reconstructed by the scalar observer (actually, the scalar observers are constructed at the first stage precisely for these functionals), and

y_1 is a measured output of the system. Thus, constructing $(\nu_1 - 1)$ scalar observers for $\sigma(\lambda_j)$ at the first stage, we can reconstruct the required functional $\sigma^1 = F^1 x$ (from decomposition (4.76)) asymptotically accurately and reduce the other functionals to the form $\tilde{\sigma}^i = \tilde{F}^i x$, where $\tilde{F}^i$ has the form from (4.80), i.e., does not depend on the first subsystem, by the linear transformation (4.81).

As a result, the problem reduces to the construction of a functional observer for the functional

$$\tilde{\sigma} = \tilde{F}x = \begin{pmatrix} \tilde{F}^2 \\ \tilde{F}^3 \\ \vdots \\ \tilde{F}^p \end{pmatrix} x, \quad \tilde{\sigma} \in \mathbb{R}^{p-1},$$

whose dimension is lowered by one.

In this case, the functional is defined not by the whole system (4.71) but only by its reduced part, without the first subsystem. For this part the observability index is equal to the dimension of the maximal one of the remaining subsystems, i.e., ν_2; the dimension of the output is also lowered by one:

$$\tilde{y} = \begin{pmatrix} y_2 \\ \vdots \\ y_l \end{pmatrix}.$$

For solving the reduced problem we can use the scheme of construction of the functional observer proposed above with the employment of scalar observers. At the second stage we shall construct $(\nu_2 - 1)$ scalar observers for the functionals

$$\sigma(\lambda_j), \quad j = \nu_1, \dots, \nu_1 + \nu_2 - 2, \quad \text{where} \quad \lambda_i \neq \lambda_j \quad \text{for} \quad i \neq j.$$

Then the dimension of the problem can be lowered again. Continuing constructions by induction, we obtain a system from

$$k(p) = (\nu_1 - 1) + (\nu_2 - 1) + \cdots + (\nu_p - 1)$$

scalar observers (if $p \leq l$) according to which (together with the outputs y_i) we decompose the required p scalar functionals. If $1 < p < l$, then

$$(\nu_1 - 1) < k(p) < \sum_{i=1}^{l} (\nu_i - 1) = n - l,$$

i.e., the dimension of the constructed functional observer exceeds $(\nu - 1) = (\nu_1 - 1)$ which is the dimension of the functional observer for the scalar functional obtained in [87] but lower than $(n - l)$ which is the dimension of the Luenberger observer for the full-phase vector.

If $p \geq l$, then, in order to reconstruct the functional, we can use the Luenberger observer for the full-phase vector.

Thus, we have proved the following theorem.

Theorem 4.46. *Suppose that the dynamical system (4.69) be observable and is reduced to the Luenberger canonical form and let $\nu_1, \dots, \nu_l$ be the Kronecker indices of this system arranged according to the nonincrease. Then the functional $\sigma = Fx$, $\sigma \in \mathbb{R}^p$, can be reconstructed by an observer of order $k(p)$:*

$$k(p) = \sum_{i=1}^{\min(p,l)} (\nu_i - 1). \tag{4.82}$$

In this case, for any set $\Lambda = (\lambda_1, \dots, \lambda_{k(p)})$ such that $\lambda_i \neq \lambda_j$ for $i \neq j$, $\lambda_i < 0$, there exists an arbitrarily close to it set $\tilde{\Lambda}$ such that $\tilde{\Lambda}$ is the spectrum of this observer.

In a certain sense, estimate (4.82) is the best possible, namely, for any $p \geq 1$ there exist functionals $\sigma \in \mathbb{R}^p$ (i.e., there exist matrices $F \in \mathbb{R}^{p\times n}$) such that they cannot be reconstructed by an observer of an order lower than $k(p)$.

Example 4.47. As an example of such an observer we can consider $\sigma = Fx$, where the matrix F has the form

$$F = \begin{pmatrix} e_{\nu_1} & 0 & \dots & 0 \\ 0 & e_{\nu_2} & \dots & 0 \\ \vdots & \vdots & \vdots & \vdots \\ 0 & 0 & \dots & e_{\nu_l} \end{pmatrix} \in \mathbb{R}^{p\times n}, \quad e_{\nu_i} = (1, 0, \dots, 0) \in \mathbb{R}^{\nu_i}. \tag{4.83}$$

The functional σ consists of p scalar functionals $\sigma^i = e_{\nu_i} x^i$ each of which corresponds to the ith subsystem from (4.71). Therefore, in this case the problem decomposes into p independent subproblems for separate scalar subsystems and also scalar functionals.

Each problem of this kind is solved independently of other problems by an observer of order $(\nu_i \quad 1)$, the total dimension of the observer is equal to $k(p)$.

Note that the dimension of the observer cannot be lowered even by the choice of a special set Λ.

Indeed, as was shown above, the scalar functional $\sigma = Fx$ defined in the canonical basis by the row $F = (f_1, \dots, f_n)$ can be reconstructed by an observer of order $k < n - 1$ if and only if among the solutions of the linear system

$$\begin{pmatrix} f_1 & f_2 & \cdots & f_k \\ \vdots & \vdots & & \vdots \\ f_{n-k-1} & f_{n-k} & \cdots & f_{n-2} \end{pmatrix} \begin{pmatrix} l_1 \\ \vdots \\ l_k \end{pmatrix} = - \begin{pmatrix} f_{k+1} \\ \vdots \\ f_{n-1} \end{pmatrix}$$

there is a Hurwitz column $l = \begin{pmatrix} l_1 \\ \vdots \\ l_k \end{pmatrix}$.

Let us consider a row $F = (1, 0, \dots, 0)$. For this row, the indicated system assumes the form

$$\begin{pmatrix} 1 & 0 & \cdots & 0 \\ \vdots & \vdots & & \vdots \\ 0 & 0 & \cdots & 0 \end{pmatrix} \begin{pmatrix} l_1 \\ \vdots \\ l_k \end{pmatrix} = - \begin{pmatrix} 0 \\ \vdots \\ 0 \end{pmatrix}.$$

The solution of this equation for all k has the form

$$l_1 = 0, l_2, l_3, \dots, l_k \text{ are arbitrary.}$$

The column l corresponds to the polynomial $p(s) = s^k + l_k s^{k-1} + \dots + l_2 s$ one root of which is necessarily zero, i.e., the polynomial $p(s)$ is not a Hurwitz polynomial. Thus, for any $k < n - 1$ there does not exist an observer which would reconstruct the given functional.

Let us now return to the vector functional (4.83) for the vector system. As was indicated above, in this case the problem decomposes into p independent problems for the scalar functionals $\sigma^i = e_{\nu_i} x^i$ for scalar subsystems. For each subsystem there does not exist an observer of the lowered order $k < \nu_i - 1$. Consequently, functional (4.83) cannot be reconstructed by an observer of an order lower than $k(p)$.

However, functional (4.83) is rather an exception. In the majority of cases (i.e., for all functionals $\sigma = Fx$, $F \in \mathbb{R}^{p \times n}$, $p < n$, except for a set of measure zero, i.e., except for functionals whose matrices F belong to certain manifolds in the space $\mathbb{R}^{p \times n}$) the estimate $k(p)$ can be perfected. Moreover, in this case the rate of convergence of the estimate is assigned arbitrarily as before.

Let us consider the following example in order to illustrate this possibility.

Example 4.48. Suppose that we are given a system with two outputs and the Kronecker indices $\nu_1 = 3$, $\nu_2 = 2$. The dimension of the system is $n = 5$ in this case. We have to reconstruct the two-dimensional functional $\sigma = Fx \in \mathbb{R}^2$, where

$$F = \left(\begin{array}{ccc|cc} f_1^1 & f_2^1 & f_3^1 & f_4^1 & f_5^1 \\ \hline f_1^2 & f_2^2 & f_3^2 & f_4^2 & f_5^2 \end{array}\right) = \left(\begin{array}{c|c} F_1^1 & F_2^1 \\ \hline F_1^2 & F_2^2 \end{array}\right) = \left(\begin{array}{c} F^1 \\ \hline F^2 \end{array}\right).$$

Then we have

$$k(p) = (\nu_1 - 1) + (\nu_2 - 1) = 3.$$

We define the spectrum of the observer by the set

$$\Lambda = \{\lambda_1, \lambda_2, \lambda_3\}.$$

In accordance with the method of scalar observers, we have to find the coefficients γ_i and $\bar{\gamma}_j$ in the decomposition

$$F^1 = (\gamma_1(1, \lambda_1, \lambda_1^2) \vdots \gamma_2(1, \lambda_1)) + (\gamma_3(1, \lambda_2, \lambda_2^2) \vdots \gamma_4(1, \lambda_2))$$
$$+(\bar{\gamma}_1(0, 0, 1) \vdots \bar{\gamma}_2(0, 1)), \qquad (4.84)$$

where the first and second terms correspond to the scalar observers (i.e., observers for the scalar functionals $\sigma(\lambda_1)$ and $\sigma(\lambda_2)$ corresponding to the eigenvalues λ_1 and λ_2) and the last term corresponds to the outputs of the system y_1 and y_2.

In accordance with the scheme of lowering the problem described above, we shall consider for F^2 a decomposition of the form

$$F^2 = \alpha_1(\gamma_1(1, \lambda_1, \lambda_1^2); \gamma_2(1, \lambda_1)) + \alpha_2(\gamma_3(1, \lambda_2, \lambda_2^2); \gamma_4(1, \lambda_2))$$
$$+((0, 0, 0); \gamma_5(1, \lambda_3)) + (\bar{\gamma}_3(0, 0, 1); \bar{\gamma}_4(0, 1)). \qquad (4.85)$$

In this decomposition the first two terms correspond to the functionals $\sigma(\lambda_1)$ and $\sigma(\lambda_2)$ reconstructed at the first stage, the third term corresponds to the additional scalar observer (for a certain functional $\sigma(\lambda_3)$) added at the second stage of the procedure. The last term corresponds to the measured outputs y_1 and y_2

Thus, in order to obtain the whole solution of the problem, in accordance with the algorithm described above, we have to construct three scalar observers and also find the coefficients α_i, γ_i and $\bar{\gamma}_i$ in decompositions (4.84), (4.85).

Note that for the fixed set $\lambda_1, \lambda_2, \lambda_3$ these decompositions define a nonlinear system consisting of ten equations (for the coordinates f_i^j, $i = 1, \dots, 5$, $j = 1, 2$) relative to eleven variables $\gamma_1, \dots, \gamma_5$; $\bar{\gamma}_1, \dots, \bar{\gamma}_4, \alpha_1, \alpha_2$. Thus, the number of unknowns exceeds the number of equations, and this makes it possible to cancel to number of scalar observers by unity (for almost all functionals). In order to indicate this possibility, we shall carry out a more detailed analysis of decompositions (4.84), (4.85).

We shall begin with considering decomposition (4.84). If a part of the vector F^1 corresponding to the first subsystem of dimension $\nu_1 = 3$, i.e., the vector (f_1^1, f_2^1, f_3^1), is not identically zero and noncollinear with the vector $\bar{C}_1 = (0, 0, 1)$, then, as follows from Statement 4.44, the coefficients γ_1, γ_3 and $\bar{\gamma}_1$ are nonzero and are uniquely defined. For this purpose, it is required that the following condition should be fulfilled:

$$(f_1^1)^2 + (f_2^1)^2 \neq 0. \qquad (4.86)$$

Let the vector (f_1^2, f_2^2, f_3^2), i.e., the part of the vector F^2 corresponding to the first subsystems, is not identically zero either and noncollinear with $\bar{C}_1$, for which purpose it is required that

$$(f_1^2)^2 + (f_2^2)^2 \neq 0. \qquad (4.87)$$

Then this subvector is uniquely decomposed in accordance with the basis

$$(\gamma_1(1, \lambda_1, \lambda_1^2)), \quad (\gamma_3(1, \lambda_2, \lambda_2^2)), \quad ((0, 0, 1)),$$

i.e., the coefficients α_1, α_2 and $\bar{\gamma}_3$ are uniquely defined (moreover, in accordance with Statement 4.44, these coefficients are nonzero for almost all λ_1, λ_2).

Thus, from the six equations (for f_1^1, f_2^1, f_3^1, f_1^2, f_2^2 and f_3^2) six coefficients γ_1, γ_3, $\bar{\gamma}_1$, $\bar{\gamma}_3$, α_1, α_2 are uniquely defined. Let us consider the remaining equations

$$\begin{cases} \gamma_2 + \gamma_4 = f_4^1 \\ \lambda_1\gamma_2 + \lambda_2\gamma_4 + \bar{\gamma}_2 = f_5^1 \\ \gamma_2\alpha_1 + \gamma_4\alpha_2 + \gamma_5 = f_4^2 \\ \gamma_2\alpha_1\lambda_1 + \gamma_4\alpha_2\lambda_2 + \lambda_3\gamma_5 + \bar{\gamma}_4 = f_5^2 \end{cases} \tag{4.88}$$

as a system consisting of four linear equations relative to the five remaining coefficients γ_2, γ_3, $\bar{\gamma}_2$, $\bar{\gamma}_4$, γ_5. It is obvious that by choosing $\bar{\gamma}_2$, $\bar{\gamma}_4$ and γ_5 we can satisfy the last three relations for any values of the parameters. The first equation is easily satisfied by the choice of γ_2, γ_4. Thus, for all f_j^i system (4.88) is solvable.

However, the number of variables is redundant. Let us use this fact in order to lower the order of the observer. For our purpose we refuse the employment of the additional observer $\sigma(\lambda_3)$. In terms of decompositions (4.84), (4.85) this means that $\gamma_5 = 0$. Then system (4.88) turns into a system of four equations relative to four unknowns. This system has the form

$$\begin{pmatrix} 1 & 1 & 0 & 0 \\ \lambda_1 & \lambda_2 & 1 & 0 \\ \alpha_1 & \alpha_2 & 0 & 0 \\ \lambda_1\alpha_1 & \lambda_2\alpha_2 & 0 & 1 \end{pmatrix} \begin{pmatrix} \gamma_2 \\ \gamma_4 \\ \bar{\gamma}_2 \\ \bar{\gamma}_4 \end{pmatrix} = \begin{pmatrix} f_4^1 \\ f_5^1 \\ f_4^2 \\ f_5^2 \end{pmatrix}.$$

This system can be uniquely solved if and only if the matrix of this system is nondegenerate. This condition is fulfilled if and only if

$$\det \begin{pmatrix} 1 & 1 \\ \alpha_1 & \alpha_2 \end{pmatrix} = \alpha_1 - \alpha_2 \neq 0. \tag{4.89}$$

The condition $\alpha_1 = \alpha_2$ means that in decomposition (4.84), (4.85) for the vectors F^1 and F^2 the first components $(f_1^1\ f_2^1)$ and $(f_1^2\ f_2^2)$ are decomposed with respect to one and the same basis with proportional coefficients, i.e., these subvectors are collinear. This means that condition (4.89) is fulfilled if and only if

$$\det \begin{pmatrix} f_1^1 & f_2^1 \\ f_1^2 & f_2^2 \end{pmatrix} = f_1^1 f_2^2 - f_2^1 f_1^2 \neq 0. \tag{4.90}$$

Note that if condition (4.90) is fulfilled, then conditions (4.86), (4.87) are also fulfilled.

Thus, if the functional

$$\sigma = \begin{pmatrix} \sigma_1 \\ \sigma_2 \end{pmatrix} = \begin{pmatrix} F^1 x \\ F^2 x \end{pmatrix}$$

satisfies condition (4.90), then we can construct for it a functional observer of order $2 < k(p) = 3$ (i.e., an observer based on two scalar observers).

Note that the matrix F from Example 4.47 does not satisfy condition (4.90). However, it is easy to see that among the matrices $F \in \mathbb{R}^{2\times 5}$ the set of all matrices which do not satisfy condition (4.90) forms a manifold of measure zero in the space $\mathbb{R}^{2\times 5}$, i.e., the construction of a two-dimensional observer in the case considered in this example is possible for almost all matrices F (for almost all functionals $\sigma = Fx$).

Let us generalize this approach to the case of functionals and systems of arbitrary dimension. As before, we shall consider a system given in the Luenberger canonical form the Kronecker indices for which are arranged in accordance with the nonincrease $\nu_1 \geq \nu_2 \geq \cdots \geq \nu_l$. The functional $\sigma = Fx \in \mathbb{R}^p$ is defined in this basis by the matrix F:

$$F = \begin{pmatrix} F^1 \\ F^2 \\ \vdots \\ F^p \end{pmatrix} = \begin{pmatrix} f_1^1 & f_2^1 & \dots & f_n^1 \\ f_1^2 & f_2^2 & \dots & f_n^2 \\ \vdots & \vdots & \vdots & \vdots \\ f_1^p & f_2^p & \dots & f_n^p \end{pmatrix} = \begin{pmatrix} F_1^1 & \dots & F_l^1 \\ F_1^2 & \dots & F_l^2 \\ \vdots & \vdots & \vdots \\ F_1^p & \dots & F_l^p \end{pmatrix}, \tag{4.91}$$

where F^i are rows of F, F_j^i is the part of the ith row corresponding to the jth subsystem, $F_j^i \in \mathbb{R}^{1\times \nu_j}$.

Note that the last coordinates in the subvectors F_j^i correspond to the measured outputs of the jth subsystems y_j for which observers may not be constructed. Therefore, to simplify the computations, we shall consider vectors $\bar{F}_j^i$ shortened by one row:

$$\bar{F}_j^i \in \mathbb{R}^{1\times(\nu_j-1)}, \quad F_j^i = (\bar{F}_j^i \; f_{\nu_1+\dots+\nu_j}^i).$$

Alongside with them we shall consider shortened basis vectors

$$\bar{F}_j(\lambda) = (1, \lambda, \lambda^2, \dots, \lambda^{\nu_j-2}) \in \mathbb{R}^{1\times(\nu_j-1)}, \quad F_j(\lambda) = (F_j(\lambda)\; \lambda^{\nu_j-1}).$$

We shall construct a functional observer of order k for which purpose, in accordance with the method of scalar observers, we shall choose a set

$$\Lambda = \{\lambda_1, \dots, \lambda_k\}, \quad \lambda_i \neq \lambda_j \quad \text{for} \quad i \neq j, \; \lambda_i \leq -\Delta < 0, \tag{4.92}$$

where Λ is the defined rate of convergence of the observer. Then, in order to solve the problem, we have to find the coefficients γ_{ij} and α_j^i in the decompositions

$$\begin{cases} (\bar{F}_1^1, \bar{F}_2^1, \dots, \bar{F}_l^1) = \sum_{j=1}^{k} \alpha_j^1 (\gamma_{1j} \bar{F}_1(\lambda_j), \gamma_{2j} \bar{F}_2(\lambda_j), \dots, \gamma_{lj} \bar{F}_l(\lambda_j)), \\ (\bar{F}_1^2, \bar{F}_2^2, \dots, \bar{F}_l^2) = \sum_{j=1}^{k} \alpha_j^2 (\gamma_{1j} \bar{F}_1(\lambda_j), \gamma_{2j} \bar{F}_2(\lambda_j), \dots, \gamma_{lj} \bar{F}_l(\lambda_j)), \\ \vdots \\ (\bar{F}_1^p, \bar{F}_2^p, \dots, \bar{F}_l^p) = \sum_{j=1}^{k} \alpha_j^p (\gamma_{1j} \bar{F}_1(\lambda_j), \gamma_{2j} \bar{F}_2(\lambda_j), \dots, \gamma_{lj} \bar{F}_l(\lambda_j)). \end{cases} \tag{4.93}$$

Here the vectors $(\gamma_{1j}\bar{F}_1(\lambda_j),\ \gamma_{2j}\bar{F}_2(\lambda_j),\ \ldots,\ \gamma_{lj}\bar{F}_l(\lambda_j))$ correspond to the scalar functionals $\sigma(\lambda_j)$ each of which can be reconstructed by the scalar observer.

Conditions (4.93) represent a nonlinear system consisting of $p(n-l)$ equations (p rows in each of which we consider the decomposition of the vector $(\bar{F}_j^i, \ldots, \bar{F}_l^i) \in \mathbb{R}^{1\times(n-l)}$ of shortened vectors $\bar{F}_j^i$). We have to determine coefficients α_j^i (they are $(p\times k)$ in number) and coefficients γ_{ij} (they are $(l\times k)$ in number). The total number of unknowns is $(p+l)k$.

We shall describe the algorithm of constructing a set of scalar observers for the vector functional ($p \geq 2$) which will make it possible to lower, as compared to $k(p)$, the upper estimate by the dimension of the functional observer with the defined rate of convergence for almost all functionals. Let us carry out a step-by-step construction of the set of functionals $\sigma(\lambda_j)$. We shall successively extend the set of eigenvalues λ_j which satisfy conditions (4.92). We shall make l steps corresponding to l subsystems in decomposition (4.71). At each step we shall add k_i vectors (and k_i eigenvalues λ_j).

As an illustration we shall consider in parallel an example of application of the algorithm.

Example 4.49. Consider a system of order $n = 8$ with the number of outputs $l = 3$ and with the set of Kronecker indices

$$\nu_1 = 3, \quad \nu_2 = 3, \quad \nu_3 = 2.$$

In this case, the functional of dimension $p = 2$ is reconstructed, i.e., $\sigma = Fx$, $F \in \mathbb{R}^{2\times 8}$:

$$F = \begin{pmatrix} 2 & -3 & 4 & | & 1 & -2 & 12 & | & 1 & 5 \\ - & - & - & - & - & - & - & - & - & - \\ 1 & -3 & 5 & | & -9 & 17 & -10 & | & 11 & 2 \end{pmatrix}.$$

Let us return to the general scheme of constructing observers.

Step 1. We choose the first $k_1 = \nu_1 - 1$ eigenvalues $\lambda_1, \ldots, \lambda_{k_1}$ from condition (4.92). By virtue of Statement 4.44, in the decomposition of the vector $\bar{F}_1^1$ (if it is nonzero, which fact is assumed in the sequel) all coefficients in the decomposition with respect to the basis $\bar{F}_1(\lambda_1), \ldots, \bar{F}_k(\lambda_1)$ are uniquely determined and are nonzero (for almost any set λ_j). Thus the condition

$$\|\bar{F}_1^1\|^2 = (f_1^1)^2 + (f_2^1)^2 + \cdots + (f_{\nu_1-1}^1)^2 \neq 0$$

should be fulfilled. Therefore we set

$$\gamma_{1j} = 1, \quad j = 1, \ldots, k_1, \tag{4.94}$$

since these coefficients in the decomposition can be normalized by the choice of α_j^1.

For the future eigenvalues (which will be added at the next stages) we set

$$\gamma_{1j} = 0, \quad j > k_1. \tag{4.95}$$

With the indicated choice of γ_{1j} we shall consider the parts of equations (4.93) corresponding to the first subsystem:

$$\bar{F}_1^i = \sum_{j=1}^{k_1} \alpha_j^i \bar{F}_1(\lambda_j) = \left(\alpha_1^i, \dots, \alpha_{k_1}^i\right) \begin{pmatrix} 1 & \lambda_1 & \dots & (\lambda_1)^{k_1-1} \\ \dots & \dots & \dots & \dots \\ 1 & \lambda_{k_1} & \dots & (\lambda_{k_1})^{k_1-1} \end{pmatrix}, \tag{4.96}$$
$$i = 1, \dots, p.$$

Then the unknown coefficients α_j^i for $i = 1, \dots, p$, $j - 1, \dots, k_1$, will bc determined from (4.96) uniquely since the matrix of the linear system (4.96) relative to these unknowns is a Vandermonde matrix.

Thus, at the first stage we determine γ_{1j} for all j (the "future" ones inclusive) as well as α_j^i for $j = 1, \dots, \nu_1 - 1$, $i = 1, \dots, p$.

Example 4.49 (continued). Step 1 for the example. For the system considered in the example we have $k_1 = \nu_1 - 1 = 2$. Consequently, at the first step we construct two scalar observers. We choose $\lambda_1 = -1$, $\lambda_2 = -2$ and obtain $\gamma_{11} = 1$, $\gamma_{12} = 1$, $\gamma_{1j} = 0$ for $j > 2$:

$$\bar{F}_1(\lambda_1) = (1, -1)$$
$$\bar{F}_2(\lambda_2) = (1, -2).$$

In order to determine α_j^i, $j = 1, 2$, $i = 1, 2$, we obtain a linear system of equations

$$\begin{cases} (2, -3) = \alpha_1^1(1, -1) + \alpha_2^1(1, -2) \\ (1, -3) = \alpha_1^2(1, -1) + \alpha_2^2(1, -2), \end{cases}$$

whence we find that

$$\alpha_1^1 = 1, \quad \alpha_2^1 = 1, \quad \alpha_1^2 = -1, \quad \alpha_2^2 = 2.$$

Let us return to the general scheme.

Step 2. At the second stage we add k_2 eigenvalues $\lambda_{k_1+1}, \dots, \lambda_{k_1+k_2}$ which satisfy condition (4.92). They will be associated with k_2 scalar observers which reconstruct the functionals $\sigma(\lambda_{k_1+1}), \dots, \sigma(\lambda_{k_1+k_2})$. By analogy with the first step we set

$$\gamma_{2j} = 1, \quad j = k_1 + 1, \dots, k_1 + k_2, \quad \gamma_{2j} = 0, \quad j > k_1 + k_2.$$

The number of added functionals k_2 should be determined, for which purpose, by analogy with the first step, we consider the part of decomposition (4.93) corresponding to the second subsystem

$$\bar{F}_2^i = \sum_{j=1}^{k_1} \alpha_j^i \gamma_{2j} \bar{F}_2(\lambda_j) + \sum_{j=k_1+1}^{k_1+k_2} \alpha_j^i \bar{F}_2(\lambda_j), \tag{4.97}$$

where $\bar{F}_2^i$, $\bar{F}_2(\lambda_j) \in \mathbb{R}^{1\times(\nu_2-1)}$, the coefficients α_j^i for $j = 1, \ldots, k_1$ (i.e., in the first sum), were determined at the preceding stage. Thus (4.97) is a system of linear equations relative to the parameters α_j^i for $j = k_1 + 1, \ldots, k_1 + k_2$, $i = 1, \ldots, p$, and γ_{2j} for $j = 1, \ldots, k_1$. Note that although (4.97) is, as the preceding decomposition (4.93), a nonlinear system relative to the total set of variables α_j^i and γ_{2j}, in the first group of terms α_j^i were determined at the first step and in (4.97) are known coefficients of the linear system.

System (4.97) consists of $p(\nu_2 - 1)$ equations, and we have to find $k_1 + pk_2$ unknowns.

In the general case (i.e., for almost all matrices F and almost all λ_j) for the system to be solvable the number of variables should be not smaller than the number of equations. Hence we get an estimate of k_2 which is the number of scalar observers added at the second stage:

$$k_1 + pk_2 \geq p(\nu_2 - 1).$$

Solving this inequality, we obtain

$$k_2 \geq (\nu_2 - 1) - \frac{k_1}{p}.$$

We choose the minimal nonnegative integer k_2 satisfying this condition:

$$k_2 = \max\left\{(\nu_2 - 1) - \left[\frac{k_1}{p}\right], 0\right\}, \tag{4.98}$$

where $[\cdot]$ is the integer part of the number.

System (4.97) has a solution if the rank of the matrix of the system coincides with the rank of the extended matrix. Under condition (4.98) (if $k_2 > 0$) for the system to be solvable it is sufficient that the matrix of the system should have a full rank.

If the system is solvable, then we find from it the remaining coefficients γ_{2j} (for $j = 1, \ldots, k_1$) and α_j^i ($j = k_1 + 1, \ldots, k_1 + k_2$, $i = 1, \ldots, p$). The latter will be used at the next stages.

Example 4.49 (continued). Step 2 for the example. We substitute into (4.98) the values $p = 2$, $\nu_2 = 3$ and $k_1 = \nu_1 - 1 = 2$ considered in the example and obtain

$$k_2 = \max\left\{(\nu_3 - 1) - \left[\frac{2}{2}\right], 0\right\} = 1.$$

Consequently, at the second step we add one eigenvalue, and then

$$k_2 = 1 < \nu_2 - 1 = 2.$$

We choose $\lambda_3 = -3$ (earlier we have chosen $\lambda_1 = -1$ and $\lambda_2 = -2$), and then

$$\begin{aligned}\bar{F}_2(\lambda_1) &= (1, -1)\\ \bar{F}_2(\lambda_2) &= (1, -2)\\ \bar{F}_2(\lambda_3) &= (1, -3).\end{aligned}$$

Let us write out the linear system of type (4.97) taking into account the values α_1^1, α_2^1, α_1^2 and α_2^2 found at the first step:

$$\begin{cases}(1, -2) = \gamma_{21}(1, -1) + \gamma_{22}(1, -2) + \alpha_3^1(1, -3)\\ (-9, 17) = \gamma_{21}(-1, 1) + \gamma_{22}(2, -4) + \alpha_3^2(1, -3),\end{cases}$$

whence we find the unique solution

$$\alpha_3^1 = 2, \quad \alpha_3^2 = -1, \quad \gamma_{21} = 2, \quad \gamma_{22} = -3.$$

Let us return to the general scheme.

Step q. Suppose that $q - 1$ steps of the algorithm are performed. We have chosen $(k_1 + k_2 + \cdots + k_{q-1})$ eigenvalues λ_j, at the preceding stages the values α_j^i, $j = 1, \ldots, q-1$, $i = 1, \ldots, p$, were determined, as well as γ_{ij}, $i = 1, \ldots, q-1$, for all j. At the next step we add k_q eigenvalues $\lambda_{k_1+\cdots+k_{q-1}+1}, \ldots, \lambda_{k_1+\cdots+k_q}$ satisfying condition (4.92). As before, we set

$$\begin{aligned}\gamma_{qj} &= 1, \quad j = k_1 + \cdots + k_{q-1} + 1, \ldots, k_1 + \cdots + k_q\\ \gamma_{qj} &= 0, \quad j > k_1 + \cdots + k_q.\end{aligned} \tag{4.99}$$

We write out the part of system (4.93) corresponding to the qth subsystem

$$\bar{F}_q^i = \sum_{j=1}^{k_1+\cdots+k_{q-1}} \alpha_j^i \gamma_{qj} \bar{F}_q(\lambda_j) + \sum_{j=k_1+\cdots+k_{q-1}+1}^{k_1+\cdots+k_q} \alpha_j^i \bar{F}_q(\lambda_j), \quad i = 1, \ldots, p. \tag{4.100}$$

Here, as before, the first group of terms corresponds to the eigenvalues λ_j chosen at the preceding steps, in this group α_j^i were determined earlier and γ_{qj}, $j = 1, \ldots, k_1 + \cdots + k_{q-1}$, should be determined (the remaining γ_{qj} can be found from (4.99)). In the second group of terms α_j^i should be determined for $j = k_1 + \cdots + k_{q-1} + 1, \ldots, k_1 + \cdots + k_q$ (i.e., for k_q new eigenvalues) and $i = 1, \ldots, p$. Thus (4.100) is a system of $p(\nu_q - 1)$ equations (since $\bar{F}_q^i$, $\bar{F}_q(\lambda_j) \in \mathbb{R}^{1\times(\nu_q-1)}$), from which we have to find $(k_q p + (k_1 + k_2 + \cdots + k_{q-1}))$ unknowns (α_j^i and γ_{qj}, respectively).

For this system to be solvable (for almost all functionals and sets of λ_j) the number of variables must be not smaller than the number of equations, i.e.,

$$k_q p + (k_1 + k_2 + \cdots + k_{q-1}) \geq p(\nu_q - 1).$$

Hence we obtain an estimate for k_p which is the number of scalar observers added at the next step

$$k_q \geq (\nu_q - 1) - \frac{(k_1 + k_2 + \cdots + k_{q-1})}{p}.$$

The minimal nonnegative integer satisfying this condition is

$$k_q = \max\left\{(\nu_q - 1) - \left[\frac{(k_1 + k_2 + \cdots + k_{q-1})}{p}\right], 0\right\},$$

where, as before, $[\cdot]$ denotes the integer part of the number.

Example 4.49 (continued). Step 3 for the example. We substitute the determined values $p = 2$, $\nu_3 = 2$, $k_2 = 1$, and $k_1 = 2$ and obtain

$$k_3 = \max\left\{(2 - 1) - \left[\frac{1+2}{2}\right], 0\right\} = \max\{0, 0\} = 0.$$

Thus, at the third step no new eigenvalues are added (and, consequently, no scalar observers). In this case, the second group of terms will be absent in system (4.100) and, with due account of α_j^i found earlier, the equations will assume the form

$$\begin{cases} 1 = \gamma_{31} + \gamma_{32} + 2\gamma_{33} \\ 11 = -\gamma_{31} + 2\gamma_{32} - \gamma_{33}. \end{cases}$$

This system has infinitely many solutions, in particular, one of them is

$$\gamma_{31} = 2, \quad \gamma_{32} = 5, \quad \gamma_{33} = -3.$$

Let us return to the general scheme.

The completion of the algorithm. Acting in the manner described above, we construct a set of $k^* = \sum_{i=1}^{l} k_i$ eigenvalues $\lambda_1, \ldots, \lambda_{k^*}$ and the functionals $\sigma(\lambda_i)$ corresponding to them, where

$$\begin{aligned} \sigma(\lambda_j) &= \hat{F}(\lambda_j)x \\ \hat{F}(\lambda_j) &= (\gamma_{1j} F_1(\lambda_j), \gamma_{2j} F_2(\lambda_j), \ldots, \gamma_{lj} F_l(\lambda_j)) \\ F_i(\lambda_j) &= \left(1, \lambda_j, \lambda_j^2, \ldots, \lambda_j^{\nu_i - 1}\right) \in \mathbb{R}^{1\times\nu_j} \end{aligned} \tag{4.101}$$

are functionals which are reconstructed by scalar observers. Moreover, all coefficients α_j^i are determined in the decomposition of the required functionals $\sigma^i = F^i x$, $i = 1, \dots, p$, from the functionals $\sigma(\lambda_j)$:

$$\sigma^i = \sum_{j=1}^{k^*} \alpha_j^i \sigma(\lambda_j) + \sum_{q=1}^{l} \bar{\alpha}_q^i y_q, \tag{4.102}$$

where the coefficients $\bar{\alpha}_q^i$ are uniquely determined from the linear system (4.102).

Example 4.49 (completion). In the example under consideration we have found $k_1 = 2$, $k_2 = 1$, and $k_3 = 0$ and have

$$k^* = k_1 + k_2 + k_3 = 3 < k(p) = (\nu_1 - 1) + (\nu_2 - 1) = 4.$$

We have chosen three eigenvalues $\lambda_1 = -1$, $\lambda_2 = -2$, $\lambda_3 = -3$ and the functionals $\sigma(\lambda_i) = \hat{F}(\lambda_i)x$, $i = 1, 2, 3$, corresponding to them, where

$$\hat{F}(\lambda_1) = (1(1,-1,1), -2(1,-1,1), 2(1,-1)) = (1,-1,1,-2,2,-2,2,-2)$$

$$\hat{F}(\lambda_2) = (1(1,-2,4), -3(1,-2,4), 5(1,-2)) = (1,-2,4,-3,6,-12,5,-10)$$

$$\hat{F}(\lambda_3) = (0(1,-3,9), 1(1,-3,9), 3(1,-3)) = (0,0,0,1,-3,9,-3,9).$$

Each functional $\sigma(\lambda_i)$ can be reconstructed by a scalar observer. Together with the outputs $y_j = C_j x$ $(j = 1, 2, 3)$ they give, in linear combinations, both scalar components $\sigma^i = F^i x$ of the required functional:

$$\begin{aligned} F^1 &= ((2,-3,4), (1-2,12), (1,5)) \\ &= \sum_{j=1}^{3} \alpha_j^1 \hat{F}(\lambda_j) + \bar{\alpha}_1^1((0,0,1),(0,0,0),(0,0)) \\ &\quad + \bar{\alpha}_2^1((0,0,0),(0,0,1),(0,0)) + \bar{\alpha}_3^1((0,0,0),(0,0,0),(0,1)) \end{aligned}$$

$$\begin{aligned} F^2 &= ((1,-3,5), (-9,17,-10), (11,2)) \\ &= \sum_{j=1}^{3} \alpha_j^2 \hat{F}(\lambda_j) + \bar{\alpha}_1^2((0,0,1),(0,0,0),(0,0)) \\ &\quad + \bar{\alpha}_2^2((0,0,0),(0,0,1),(0,0)) + \bar{\alpha}_3^2((0,0,0),(0,0,0),(0,1)). \end{aligned}$$

Knowing α_j^i, we find $\bar{\alpha}_j^i$ from these equations:

$$\begin{aligned} \bar{\alpha}_1^1 &= -19, & \bar{\alpha}_2^1 &= -11, & \bar{\alpha}_3^1 &= 14, \\ \bar{\alpha}_1^2 &= 7, & \bar{\alpha}_2^2 &= -8, & \bar{\alpha}_3^2 &= 2. \end{aligned}$$

Thus, the required two-dimensional functional can be reconstructed by a functional observer of order $k^* = 3$. This observer gives an exponential estimate of the required functional, and the rate of convergence of the observation error is chosen arbitrarily.

In the example that we have considered, at the third step $k_3 = 0$, i.e., at the third step we do not add new scalar observers. In the general case we have the following lemma.

Lemma 4.50. *Suppose that we are given a dynamical system of the general position of order n with l outputs reduced to the Luenberger canonical form, $\nu_1 \geq \nu_2 \geq \cdots \geq \nu_l$ are Kronecker indices arranged nonincreasingly. Suppose that $p \leq n$. Let us define the coefficients k_i for $i = 1, \dots, l$ by the relations*

$$k_1 = \nu_1 - 1$$
$$k_i = \max\left\{(\nu_i - 1) - \left[\frac{1}{p}\sum_{j=1}^{i-1} k_j\right], 0\right\}, \quad i = 2, \dots, l, \tag{4.103}$$

where $[\cdot]$ is the integer part of the number. Let $k_{i^*} = 0$ for a certain i^*. Then $k_i = 0$ for all $i > i^*$.

Proof. The statement follows from definition (4.103). Indeed, if $k_{i^*} = 0$, then we have

$$(\nu_{i^*} - 1) - \left[\frac{1}{p}\sum_{j=1}^{i^*-1} k_j\right] \leq 0.$$

Moreover, it is obvious that

$$\sum_{j=1}^{i^*-1} k_j = \sum_{j=1}^{i^*} k_j.$$

However, in that case, taking into account that $\nu_{i^*+1} \leq \nu_{i^*}$, we obtain

$$(\nu_{i^*+1}-1)-\left[\frac{1}{p}\sum_{j=1}^{i^*} k_j\right] = (\nu_{i^*+1}-1)-\left[\frac{1}{p}\sum_{j=1}^{i^*-1} k_j\right] \leq (\nu_{i^*}-1)-\left[\frac{1}{p}\sum_{j=1}^{i^*-1} k_j\right] \leq 0.$$

It follows that

$$k_{i^*+1} = \max\left\{(\nu_{i^*+1} - 1) - \left[\frac{1}{p}\sum_{j=1}^{i^*} k_j\right], 0\right\} = 0.$$

Thus, if $k_{i^*} = 0$, then $k_{i^*+1} = 0$. The lemma is proved. □

In the example we have $k_i = 0$ for $i > p$. In the general case this condition may not be fulfilled. Suppose that we are given a system for $n = 15$, $\nu_1 = \nu_2 = \nu_3 = 5$.

For $p = 2$ we have

$$k_1 = (\nu_1 - 1) = 4$$

$$k_2 = (\nu_2 - 1) - \left[\frac{k_1}{2}\right] = 4 - \left[\frac{4}{2}\right] = 2$$

$$k_3 = (\nu_3 - 1) - \left[\frac{k_1 + k_2}{2}\right] = 4 - \left[\frac{6}{2}\right] = 1 > 0.$$

Thus, although $k_i \le (\nu_i - 1)$, the sum for $k^*(p)$ may include a larger number of terms (for $p < 1$) than the sum for $k(p)$. Nevertheless, in all cases $k^*(p) \le k(p)$. To be more precise, we have the following lemma.

Lemma 4.51. *Suppose that we are given a dynamical system of the general position of order n with l outputs reduced to the canonical Luenberger form, $\nu_1 \ge \nu_2 \ge \dots \ge \nu_l$ are Kronecker indices arranged nonincreasingly. Let*

$$k^*(p) = \sum_{i=1}^{l} k_i,$$

where k_i were determined in (4.103).

Then, for any p, we have an estimate

$$k^*(p) \le k(p) = \sum_{i=1}^{\min\{p,l\}} (\nu_i - 1).$$

Proof. By construction we have $k_i \le (\nu_i - 1)$. In this case, the statement of the lemma obviously follows for $p \ge l$ (since $k^*(p) = \sum_{i=1}^{l} k_i$ and $k(p) = \sum_{i=1}^{l}(\nu_i - 1)$ in this case).

Let $p < l$. Let us prove the statement by induction. To be more precise, we shall prove that for all $1 \le q \le l$ we have an inequality

$$\sum_{i=1}^{q} k_i \le \sum_{i=1}^{p}(\nu_i - 1) = k(p).$$

For $1 \le q \le p$ this inequality is obvious. Suppose that it is satisfied at a certain step $q \ge p$. In this case we denote by t the difference

$$t = \sum_{i=1}^{p}(\nu_i - 1) - \sum_{i=1}^{q} k_i \ge 0.$$

Then for $(q + 1)$, when $k_{q+1} > 0$ (for $k_{q+1} = 0$ the statement is trivial), we have an estimate

$$\sum_{i=1}^{p}(\nu_i - 1) - \sum_{i=1}^{q+1} k_i = \sum_{i=1}^{p}(\nu_i - 1) - \sum_{i=1}^{q} k_i - k_{q+1}$$

$$
\begin{aligned}
&= t - \nu_{q+1} + 1 + \left[\frac{1}{p}\sum_{i=1}^{q} k_i\right] \\
&= \left[\frac{1}{p}\left(tp - p(\nu_{q+1} - 1) + \sum_{i=1}^{q} k_i\right)\right] \\
&= \left[\frac{1}{p}\left(tp - t + \left(t + \sum_{i=1}^{q} k_i\right) - p(\nu_{q+1} - 1)\right)\right] \\
&= \left[\frac{1}{p}\left(t(p-1) + \sum_{i=1}^{p}(\nu_i - 1) - p(\nu_{q+1} - 1)\right)\right] \\
&= \left[\frac{1}{p}\left(t(p-1) + \sum_{i=1}^{p}(\nu_i - \nu_{q+1})\right)\right] \geq 0.
\end{aligned}
$$

Thus, if the inequality is valid for q, then it is also valid for $(q+1)$. Consequently, by induction, it is also valid for $q = l$, and this is what we had to prove. □

The statement of Lemma 4.51 implies the main theorem.

Theorem 4.52. *Suppose that we are given a dynamical system of the general position of order n with l outputs reduced to the Luenberger canonical form, $\nu_1 \geq \nu_2 \geq \cdots \geq \nu_l$ are Kronecker indices arranged according to the nonincrease.*

Then, for almost all matrices $F \in \mathbb{R}^{p\times n}$ for the functional $\sigma = Fx \in \mathbb{R}^p$, for any stable real and distinctive spectrum $\Lambda = \{\lambda_1, \ldots, \lambda_{k^*}\}$, there exists an arbitrarily close to it stable real and distinctive spectrum Λ' such that for the functional σ we can construct an observer of order $k^*(p)$ with spectrum Λ'.

Proof. By construction, the algorithm of synthesis of an observer described above is applicable to the given functional if and only if the systems of linear equations appearing at every step of the algorithm have solutions. Let us write, in matrix form, the system of equations appearing at the ith step of the algorithm:

$$
\begin{aligned}
x_i M_i &= z_i \\
x_i &= \left(\gamma_{i1}, \ldots, \gamma_{ik_i'}, \alpha^1_{k_i'+1}, \ldots, \alpha^1_{k_i'+k_i}, \alpha^2_{k_i'+1}, \ldots, \alpha^p_{k_i'+k_i}\right) \\
z_i &= \left(\bar{F}_i^1, \bar{F}_i^2, \ldots, \bar{F}_i^p\right)
\end{aligned}
$$

$$M_i = \begin{pmatrix} \alpha_1^1 \bar{F}_i(\lambda_1) & \alpha_1^2 \bar{F}_i(\lambda_1) & \dots & \alpha_1^p \bar{F}_i(\lambda_1) \\ \alpha_2^1 \bar{F}_i(\lambda_2) & \alpha_2^2 \bar{F}_i(\lambda_2) & \dots & \alpha_2^p \bar{F}_i(\lambda_2) \\ \dots & \dots & \dots & \dots \\ \alpha_{k_i'}^1 \bar{F}_i(\lambda_{k_i'}) & \alpha_{k_i'}^2 \bar{F}_i(\lambda_{k_i'}) & \dots & \alpha_{k_i'}^p \bar{F}_i(\lambda_{k_i'}) \\ \bar{F}_i(\lambda_{k_i'+1}) & 0 & \dots & 0 \\ \dots & \dots & \dots & \dots \\ \bar{F}_i(\lambda_{k_i'+k_i}) & 0 & \dots & 0 \\ 0 & \bar{F}_i(\lambda_{k_i'+1}) & \dots & 0 \\ \dots & \dots & \dots & \dots \\ 0 & \bar{F}_i(\lambda_{k_i'+k_i}) & \dots & 0 \\ \dots & \dots & \dots & \dots \\ 0 & 0 & \dots & \bar{F}_i(\lambda_{k_i'+1}) \\ \dots & \dots & \dots & \dots \\ 0 & 0 & \dots & \bar{F}_i(\lambda_{k_i'+k_i}), \end{pmatrix},$$

where $k_i' = \sum_{j=1}^{i-1} k_j$ is the number of eigenvalues added at the preceding steps, k_i is the number of eigenvalues added at the ith step, $x_i \in \mathbb{R}^{1\times(k_i'+pk_i)}$ is the vector of the unknown parameters of the method for the ith step, $M_i \in \mathbb{R}^{(k_i'+pk_i)\times p(\nu_i-1)}$ is the matrix of the system in block form, $z_i \in \mathbb{R}^{1\times p(\nu_i-1)}$ is the right-hand side also written in block form, $\Lambda = \{\lambda_i\}$ is the defined spectrum of the observer. The system has a solution if

$$\operatorname{rank} M_i = \operatorname{rank} \begin{pmatrix} M_i \\ z_i \end{pmatrix}.$$

Since $k_i' + pk_i \geq p(\nu_i - 1)$ by construction, if $\operatorname{rank} M_i = p(\nu_i - 1)$, then, irrespective of z_i

$$\operatorname{rank} \begin{pmatrix} M_i \\ z_i \end{pmatrix} = p(\nu_i - 1)$$

and the system has a solution. Consequently, the condition of the fullness of the rank

$$\operatorname{rank} M_i = p(\nu_i - 1), \quad i = 1, \dots, l,$$

at every step is the sufficient condition of applicability of the algorithm.

Let us investigate the structure of the matrix M_i in greater detail. At the first step of the algorithm we have $k_1 = \nu_1 - 1$ and $k_1' = 0$. The matrix M_1 is quasidiagonal with p blocks of the form

$$\begin{pmatrix} \bar{F}_1(\lambda_1) \\ \vdots \\ \bar{F}_1(\lambda_{k_1}) \end{pmatrix}$$

on the principal diagonal. Each of these blocks of dimension $(\nu_1 - 1) \times (\nu_1 - 1)$ has full rank equal to $(\nu_1 - 1)$, and, consequently,

$$\operatorname{rank} M_1 = p(\nu_1 - 1),$$

at the first step of the algorithm the system has a solution for any functional and any set of different stable real eigenvalues. □

Let us now consider the matrix M_i for $i > 1$. We shall prove the following auxiliary statement.

Lemma 4.53. *By elementary transformations the matrix M_i, $i = 2, \dots, l$, can be reduced to the form*

$$M_i' = \begin{pmatrix} 0 & \alpha_1^1 F_i'(\lambda_1) & 0 & \alpha_1^2 F_i'(\lambda_1) & \cdots & 0 & \alpha_1^p F_i'(\lambda_1) \\ 0 & \alpha_2^1 F_i'(\lambda_2) & 0 & \alpha_2^2 F_i'(\lambda_2) & \cdots & 0 & \alpha_2^p F_i'(\lambda_2) \\ \vdots & \vdots & \vdots & \vdots & \vdots & \vdots & \vdots \\ 0 & \alpha_{k_i'}^1 F_i'(\lambda_{k_i'}) & 0 & \alpha_{k_i'}^2 F_i'(\lambda_{k_i'}) & \cdots & 0 & \alpha_{k_i'}^p F_i'(\lambda_{k_i'}) \\ I_{k_i} & 0 & \cdots & 0 & \cdots & 0 & 0 \\ 0 & 0 & I_{k_i} & 0 & \cdots & 0 & 0 \\ \vdots & \vdots & \vdots & \vdots & \vdots & \vdots & \vdots \\ 0 & 0 & 0 & 0 & \cdots & I_{k_i} & 0 \end{pmatrix},$$

where I_{k_i} is an identity matrix of dimension $k_i \times k_i$, $F_i'(\lambda_j) \in \mathbb{R}^{1\times(\nu_i-1-k_i)}$ have the form

$$F_j'(\lambda_i) = \left(1, \lambda_j, \dots, \lambda_j^{\nu_i-1-k_i-1}\right),$$

and the zero blocks have corresponding dimensions. By the transposition of columns the matrix M_i' can be reduced to the matrix

$$M_i'' = \left(\begin{array}{c|c} 0 & \bar{M}_i \\ \hline I & 0 \end{array}\right),$$

where I is an identity matrix of dimension $(pk_i) \times (pk_i)$ and

$$\bar{M}_i = \begin{pmatrix} \alpha_1^1 F_i'(\lambda_1) & \alpha_1^2 F_i'(\lambda_1) & \cdots & \alpha_1^p F_i'(\lambda_1) \\ \alpha_2^1 F_i'(\lambda_2) & \alpha_2^2 F_i'(\lambda_2) & \cdots & \alpha_2^p F_i'(\lambda_2) \\ \vdots & \vdots & \vdots & \vdots \\ \alpha_{k_i'}^1 F_i'(\lambda_{k_i'}) & \alpha_{k_i'}^2 F_i'(\lambda_{k_i'}) & \cdots & \alpha_{k_i'}^p F_i'(\lambda_{k_i'}) \end{pmatrix} \in \mathbb{R}^{k_i' \times p(\nu_i-1-k_i)}.$$

Proof. Here is the step-by-step list of transformations of the matrix M_i which leads to the indicated result.

(1) First we transform the part corresponding to the first scalar functional from $\sigma = Fx$.

From each column, from the second to the $(\nu_i - 1)$th, we subtract the first column multiplied by $\lambda_{k_i'+1}^j$, where j is the number of the column being transformed.

Then in the $(k_i' + 1)$th row of the first column there will be a unity and the other elements will remain zero. We shall carry out a similar operation for the parts corresponding to the other rows of F, from the second to the pth, for which purpose we subtract from every column from the $(\nu_i + 1)$th to the $(2\nu_i - 2)$th the column ν_i multiplied by $\lambda_{k_j'+1}^{j-\nu_i}$, where j is the number of the transformed column, and so on. Then, in all rows corresponding to $\lambda_{k_j'+1}$, the respective places will be occupied by unities (in the first column in the first row, in the ν_ith column in the second row, and so on), and the other elements will be zero. Consequently, subtracting these rows multiplied by the corresponding coefficients from the remaining rows, we can achieve a situation where unities will be placed at the intersection of these rows with columns with the numbers $1, (\nu_i - 1) + 1 = \nu_i, 2(\nu_i - 1) + 1 = 2\nu_i - 1, \dots, (p-1)(\nu_i - 1) + 1$ and the other elements of the indicated columns will be zero.

(2) We divide all rows of the matrix, except those which correspond to $\lambda_{k_j'+1}$, by $(\lambda_j - \lambda_{k_j'+1})$, where the transformed row corresponds to λ_j.

(3) For the part which corresponds to the first row of F we successively subtract from the $(\nu_i - 1)$th column the $(\nu_i - 2)$th column multiplied by $\lambda_{k_i'+1}$, then subtract from the $(\nu_i - 2)$th column the $(\nu_i - 3)$th column, and so on, and finally subtract the second column from the third one. We shall carry out similar operations for the parts corresponding to the rows of F from the second row to the pth one. Note that now no one of the elements of the transformed matrix depends on $\lambda_{k_i'+1}$, and only zeros or unities are in the rows corresponding to this eigenvalue.

(4) We remove from the matrix columns with numbers $1, \nu_i, 2(\nu_i - 1), \dots, (p-1)(\nu_i - 1) + 1$ and rows corresponding to $\lambda_{k_i'+2}$. It is easy to see that the remaining matrix has the form similar to the original form of M_i, only now the "new first eigenvalue" is $\lambda_{k_i'+2}$.

Consequently, for the remaining matrix we can carry out the above-indicated operations (1)–(3) with the replacement of $\lambda_{k_i'+1}$ by $\lambda_{k_i'+2}$ with a correction concerning the numbers of columns. As a result we shall get rid of the dependence on $\lambda_{k_i'+2}$.

Acting by analogy, we shall remove from the matrix all eigenvalues from $\lambda_{k_i'+1}$ to $\lambda_{k_i'+k_i}$ and, as a result, obtain the required representation of the matrix from the condition of the lemma.

The lemma is proved. □

Thus, for the condition

$$\operatorname{rank} M_i = \operatorname{rank} M_i' = \operatorname{rank} M_i'' = p(\nu_i - 1)$$

to be fulfilled, it is necessary and sufficient that

$$\operatorname{rank} \bar{M}_i = p(\nu_i - 1 - k_i). \tag{4.104}$$

Note that $\bar{M}_i$ depends only on the eigenvalues λ_j and coefficients α_j^m found at the preceding stages ($j \le k_i'$) and does not depend on the eigenvalues $\lambda_{k_i'+1}, \dots, \lambda_{k_i'+k_i}$

added at the ith step. Let us now prove that all unknown coefficients α_j^m and γ_{ij}, defined at the ith step, can be represented as fractional-rational functions of the elements of the matrix F and eigenvalues λ_j, $j = 1, \dots, \sum_{q=1}^{i} k_q$, under the condition that the algorithm is applicable up to the ith step inclusive.

For $i = 1$ this statement is valid since the matrix M_i is always nondegenerate and the coefficients α_1^m are uniquely defined by the equation

$$x_1 = z_1(M_1)^{-1},$$

where x_1 consists only of the coefficients α_1^m and the right-hand side depends only on $\lambda_1, \dots, \lambda_{k_1}$ and $f_1^1, \dots, f_{\nu_1-1}^p$.

Suppose that the statement is valid at the ith step. Then the condition of applicability of the algorithm at the $(i+1)$th step can be written in form (4.105)

$$\operatorname{rank} \bar{M}_{i+1} = p(\nu_{i+1} - 1 - k_{i+1}),$$

where the matrix $\bar{M}_{i+1}$ depends on the elements of the matrix F, the eigenvalues added at the steps up to the ith inclusive, and the coefficients defined at these steps.

Moreover, by assumption, all these coefficients can be represented as fractional-rational functions of the indicated kind. And then the condition of the fullness of the rank of the matrix $\bar{M}_{i+1}$ can be written as an inequality to zero of the sum of squares of its minors of order $p(\nu_{i+1} - 1 - k_{i+1})$, i.e., as fractional-rational functions

$$\frac{R_i(\lambda_1, \dots, \lambda_{k_i'+1}, f_1^1, \dots, f_{n-1}^p)}{T_i(\lambda_1, \dots, \lambda_{k_i'+1}, f_1^1, \dots, f_{n-1}^p)} \neq 0, \tag{4.105}$$

where R_i and T_i are polynomials of the indicated variables. Let this condition be fulfilled and suppose that the algorithm is applicable at the ith step. Then the equation

$$x_{i+1} M_{i+1} = z_{i+1}$$

is solvable and the elements of the solution x_{i+1} can be represented as fractional-rational functions of the elements M_{i+1} (depend on λ_j added at the steps up to the $(i+1)$th inclusive and on the coefficients found at the steps up to the ith inclusive). This representation can be not unique if the matrix M_{i+1} is not square.

Thus, all coefficients defined at the $(i+1)$th step are represented as fractional-rational functions of the variables which are themselves fractional-rational functions of λ_j and elements of F. Hence we have the proof of the statement.

Thus, if condition (4.106) is fulfilled after the ith step, the algorithm is applicable at the $(i+1)$th step. Consequently, if condition (4.106) is fulfilled for all $i = 1, \dots, (l-1)$, then the algorithm is applicable at all steps, and there exists, for the functional $\sigma = Fx$, an exponential observer of order $k^*(p)$ with the spectrum $\Lambda = \{\lambda_1, \dots, \lambda_{k^*}\}$ (under the condition of stability of this spectrum).

Let us consider the numerator R_i from condition (4.106) as a polynomial of $\lambda_1, \dots, \lambda_{k_i'+1}$ with coefficients dependent on the elements of F. It can be zero on the set of nonzero measure in the space of eigenvalues Λ if and only if all elements for the monomials of λ_i are identically zero.

Equivalently, R_i is identically zero if and only if there exists a set $\Lambda^0 = \{\lambda_1^0, \dots, \lambda_{k_i'+1}^0\}$ such that $R_i = 0$ in a neighborhood of Λ^0.

Each coefficient for the monomials of $\lambda_1, \dots, \lambda_{k_i'+1}$ is a polynomial of f_q^m, of the elements of the matrix F, and, simultaneously, the equality of all these polynomials to zero defines a certain set of measure zero in the space of the matrices F. If F does not belong to this set, then, in any neighborhood of *any* set Λ^0 there exists a set of nonzero measure of sets Λ such that $R_i \neq 0$ for them.

Thus, for all $i = 1, \dots, (l-1)$ the condition of applicability of the algorithm at the $(i+1)$th step (4.106) is not fulfilled only on the set of matrices F of measure zero. The union of these sets is the set of measure zero too. Now if F does not belong to this union, then in any neighborhood of the stable real distinctive spectrum Λ there exists a real stable spectrum Λ' such that (4.106) is fulfilled for all $i = 1, \dots, (l-1)$. Hence we have the statement of the theorem.

Thus, when we solve the problem of synthesis of a functional observer, we can almost always use an observer of order $k^*(p) \leq k(p)$. The set of matrices $F \in \mathbb{R}^{p \times n}$ for which it is impossible to construct such an observer is a set of measure zero in the space $\mathbb{R}^{p \times n}$.

Conclusion

In Chap. 4 we gave conditions of existence and the algorithms for synthesizing functional observers for linear stationary fully determined systems for different cases, namely, scalar and vector output, scalar and vector functional.

The authors of [87, 93] show that the functional $\sigma = Fx \in \mathbb{R}^p$ can be reconstructed by an observer of order $\nu - 1$ (where ν is the observability index of the pair $\{C, A\}$) with any preassigned rate of convergence.

In this chapter we describe new approaches to the synthesis of functional observers of the given order k ($k < \nu - 1$) which were proposed for the first time in [26–28]. Two methods are proposed for solving the problem: the method of pseudoinputs and the method of scalar observers. Both methods allow us to obtain necessary and sufficient conditions for the existence of functional observers of order k. For the scalar case $l = 1$, $p = 1$ these conditions are given by Theorem 4.3, for the case $l = 1$, $p > 1$ by Theorem 4.11, for the case $l > 1$, $p = 1$ by Theorem 4.22.

Proceeding from these theorems, we propose an algorithm for synthesizing observers of minimal order and also obtain lower bounds for the order of the observer.

In Sec. 4.6 we carried out an analysis of the necessary and sufficient conditions for the existence of observers of the given order obtained in the preceding sections. A number of auxiliary statements are proved.

In Sec. 4.7 we considered the problem of synthesis of observers with the defined dynamical properties (a given spectrum or a given rate of convergence). Theorem 4.52 gives the upper estimate for the dimension of these observers.

Chapter 5

Asymptotic observers for linear systems with uncertainty

In this chapter we consider a problem of constructing an observer for a linear stationary system subjected to the action of unknown disturbance.

We shall consider the statement of this problem more strictly. Suppose that we are given a dynamical system

$$\begin{cases} \dot{x} = Ax + Bu + Df \\ y = Cx, \end{cases} \tag{5.1}$$

where $A \in \mathbb{R}^{n\times n}$, $B \in \mathbb{R}^{n\times k}$, $D \in \mathbb{R}^{n\times m}$, $C \in \mathbb{R}^{l\times n}$ are known constant matrices, $u(t) \in \mathbb{R}^k$ and $y(t) \in \mathbb{R}^l$ are known input and output of the system respectively, $f(t,x) \in \mathbb{R}^m$ is an unknown disturbance. Using the information concerning the known input $u(t)$ and output $y(t)$, we have to construct an asymptotic estimate $\tilde{x}(t)$ of the unknown state vector $x(t) \in \mathbb{R}^n$.

Then we assume, relative to system (5.1), that the pair $\{C, A\}$ is observable. In the case of the absence of disturbance $f(t)$ the problem of constructing the estimate $\tilde{x}(t)$ was studied above, in particular, it is solved by the full-dimensional observer

$$\dot{\tilde{x}} = A\tilde{x} + Bu - L(C\tilde{x} - y), \tag{5.2}$$

where the matrix L is chosen from the condition of stability of the system in the deviations $e = \tilde{x} - x$ described by the equation

$$\dot{e} = (A - LC)e = A_L e.$$

Since the pair $\{C, A\}$ is observable, the spectrum of the matrix A_L is wholly defined by the choice of the matrix L and, consequently, the proposed full-dimensional observer solves the problem of reconstruction of the vector $x(t)$ exponentially exactly with any predefined rate of convergence.

The situation changes essentially if the system possesses an uncertainty $f(t,x)$. In this case, the system in deviations has the form

$$\dot{e} = A_L e - Df,$$

and if $f(t, x)$ does not tend to zero, then observer (5.2) does no longer give an asymptotic estimate for $x(t)$. Therefore other approaches are required for solving this problem.

The problem of synthesizing observers under the conditions of uncertainty has rich history. At present there are many methods and approaches for solving this problem. Practically all of them allow us to solve the problem under the same conditions imposed on system (5.1). We shall describe in detail two of these approaches, following [13], and then give a short review of the other methods.

5.1 Hyperoutput systems

One of the main cases considered in literature is the case of systems with the number of outputs exceeding the dimension of disturbance vector $f(t)$, the case where $l > m$. We shall call systems of this kind *hyperoutput* systems. Since we can always compensate the influence of the known input $u(t)$ in the observer, we shall assume, in what follows, for simplicity, that $u(t) \equiv 0$, i.e., consider a system

$$\begin{cases} \dot{x} = Ax + Df \\ y = Cx. \end{cases} \tag{5.3}$$

Suppose that the following assumptions are fulfilled for system (5.3).

Assumption A.1. The pair $\{C, A\}$ is observable, the pair $\{A, D\}$ is controllable, i.e., system (5.3) is in the general position.

Assumption A.2. Matrices C and D are of full rank, i.e., $\operatorname{rank} C = l$, $\operatorname{rank} D = m$.

Assumption A.3. The number of outputs exceeds the number of unknown inputs, i.e., $l > m$.

Assumption A.4. The rank condition

$$\operatorname{rank} CD = m$$

holds, i.e., the matrix $CD \in \mathbb{R}^{l \times m}$ is of full rank.

By virtue of Assumption A.4 there exists in the matrix CD a nondegenerate minor of order m. Without loss of generality, we can assume that it is in the first m rows of the matrix CD (we can always achieve this by renumbering the outputs).

Let C_i $(i = 1, \dots, l)$ be rows of the matrix C. Then

$$C = \begin{pmatrix} C' \\ C'' \end{pmatrix}, \quad C' = \begin{pmatrix} C_1 \\ \vdots \\ C_m \end{pmatrix} \in \mathbb{R}^{m \times n}, \quad C'' = \begin{pmatrix} C_{m+1} \\ \vdots \\ C_l \end{pmatrix} \in \mathbb{R}^{(l-m) \times n}.$$

From the assumptions made above it follows that the principal minor of the matrix

$$CD = \begin{pmatrix} C'D \\ C''D \end{pmatrix}$$

is nondegenerate, i.e., $\det(C'D) \neq 0$. The matrix C' corresponds to the first m components of the output and C'' to the other $(l-m)$, i.e.,

$$y = \begin{pmatrix} y' \\ y'' \end{pmatrix} = \begin{pmatrix} C'x \\ C''x \end{pmatrix}.$$

It follows from the nondegeneracy of the matrix $C'D$ that the zero dynamics of system (5.3) with respect to the output y' is of the maximal order $(n-m)$. In this case there exists a nondegenerate change of coordinates which reduces system (5.3) to the form

$$\begin{cases} \dot{x}' = A_{11}x' + A_{12}y' \\ \dot{y}' = A_{21}x' + A_{22}y' + (C'D)f, \end{cases} \tag{5.4}$$

where $x' \in \mathbb{R}^{n-m}$, A_{ij} are matrices with constant coefficients of the corresponding dimensions. Note that in the indicated representation of the system the first $(n-m)$ unknown components of the phase vector x' do not depend explicitly on the unknown disturbance f.

In order to pass to form (5.4) it suffices to take, as the first $(n-m)$ basis vectors, any basis of the subspace which is a component of the subspace spanned over the columns of the matrix D. We shall describe briefly one of the techniques of such a decomposition following [7].

Since $\det(C'D) \neq 0$, it follows that condition $\operatorname{rank} C' = m$ is fulfilled for $C' \in \mathbb{R}^{m\times n}$ and, consequently, there exists in the matrix C' a nondegenerate minor of order m. Without loss of generality we assume that it is in the last columns of the matrix $C' = (C'_{n-m}; C'_m)$, i.e., $\det C'_m \neq 0$ (here $C'_{n-m} \in \mathbb{R}^{m\times(n-m)}$, $C'_m \in \mathbb{R}^{m\times m}$). We can always achieve this by renumbering the components of the vector x. The transformation

$$x = P\begin{pmatrix} x' \\ y' \end{pmatrix}, \quad \begin{pmatrix} x' \\ y' \end{pmatrix} = P^{-1}x,$$

where

$$P = \begin{pmatrix} I_{n-m} & Q_{n-m} \\ -(C'_m)^{-1}C'_{n-m} & Q_m \end{pmatrix}, \quad P^{-1} = \begin{pmatrix} I_{n-m} - Q_{n-m}C'_{n-m} & -Q_{n-m}C'_m \\ C'_{n-m} & C'_m \end{pmatrix}$$

gives the required decomposition, the matrices $Q_{n-m} \in \mathbb{R}^{(n-m)\times m}$ and $Q_m \in \mathbb{R}^{m\times m}$ are defined by the relations

$$\begin{pmatrix} Q_{n-m} \\ Q_m \end{pmatrix} = D(C'D)^{-1},$$

$I_{n-m} \in \mathbb{R}^{(n-m)\times(n-m)}$ being an identity matrix.

After reducing the system to form (5.4), for the reconstruction of the phase vector of the system it suffices to construct an estimate for the unknown part x' of the phase vector $\begin{pmatrix} x' \\ y' \end{pmatrix}$.

If the zero dynamics of system (5.4) with respect to the output y' is stable, i.e., A_{11} is a Hurwitz matrix, then such an estimate is given, in particular, by the observer of order $(n-m)$

$$\dot{\tilde{x}}' = A_{11}\tilde{x}' + A_{12}y'. \tag{5.5}$$

In this case, the estimation error $e' = \tilde{x}' - x'$ satisfies the equation

$$\dot{e}' = A_{11}e',$$

and, consequently, the error $e' \to 0$ exponentially as $t \to \infty$. In this case the rate of convergence of the estimate is defined by the spectrum of the matrix A_{11} and cannot be changed. If the matrix A_{11} is unstable, observer (5.5) cannot be used.

However, if $l > m$, then another approach can be used for constructing an asymptotic (exponential, to be more precise) observer for x'. Note that in observer (5.5) we do not use the second part of the output $y'' = C''x \in \mathbb{R}^{l-m}$ being measured. We shall show that the use of additional information radically changes the situation.

Let us write the vector $y'' = C''x$ in the new coordinates

$$y'' = C''P^{-1}\begin{pmatrix} x' \\ y' \end{pmatrix} = C_1''x' + C_2''y', \tag{5.6}$$

where $C_1'' \in \mathbb{R}^{(l-m)\times(n-m)}$, $C_2'' \in \mathbb{R}^{(l-m)\times m}$ are matrices with constant coefficients defined by the parameters of the system.

Then, in the new coordinates $y = \tilde{C}\begin{pmatrix} x' \\ y' \end{pmatrix}$, where the matrix $\tilde{C}$ has block structure

$$\tilde{C} = CP^{-1} = \begin{pmatrix} 0 & I_m \\ C_1'' & C_2'' \end{pmatrix},$$

and, since the matrix C has full rank, we infer that the matrix C_1'' is also of full rank, i.e., $\operatorname{rank} C_1'' = l - m$.

Since y' and y'' are known vectors, we determine a new output

$$\tilde{y} = y'' - C_2''y'.$$

It follows from representation (5.6) that $\tilde{y} = C_1''x'$. Then the first equation of system (5.4) and $\tilde{y}$ can be regarded as a linear system of order $(n-m)$ with the known output y' of order $(l-m)$ and the known input $\tilde{y}$ of order m, i.e.,

$$\begin{cases} \dot{x}' = A_{11}x' + A_{12}y' \\ \tilde{y} = C_1''x'. \end{cases} \tag{5.7}$$

Note that in contrast to the original system, system (5.7) does not depend explicitly on the unknown input $f(t,x)$. Therefore, if the pair $\{C_1'', A_{11}\}$ is observable, then the problem of reconstruction of the vector x' is solved, in particular, by the full-dimensional observer of order $(n-m)$:

$$\dot{\tilde{x}}' = A_{11}\tilde{x}' + A_{12}y' - L(C_1''\tilde{x}' - \tilde{y}), \tag{5.8}$$

where the matrix $L \in \mathbb{R}^{(n-m)\times(l-m)}$ is chosen from the condition that the matrix $A_L = A_{11} - LC_1''$ is a Hurwitz matrix. In this case, the estimation error $e' = \tilde{x}' - x'$ satisfies the equation

$$\dot{e}' = A_L e'$$

and, consequently, converges to zero exponentially. Moreover, in contrast to observer (5.5), in the case of the observability of the pair $\{C_1'', A_{11}\}$, the rate of convergence is wholly determined by the choice of the matrix L and can be defined arbitrarily. In this case, the stability of the matrix A_{11} is not assumed.

Observer (5.8) also serves the problem in the case where the pair $\{C_1'', A_{11}\}$ is not observable but only reconstructible. In this case, by the choice of the matrix L we can determine a part of the spectrum of the matrix A_L, the remaining part being unchangeable and stable. The rate of convergence of the observer can be defined by the unchangeable part of the spectrum of the matrix A_L.

In addition, for system (5.7) we can construct a Luenberger observer of the lowered order $(n-m)-(l-m)=(n-l)$.

It follows from what was stated above, that the fundamental part in the construction of the observer for x' is played by the observability (reconstructibility) of the pair $\{C_1'', A_{11}\}$. To analyze it, we shall investigate the properties of the invariant zeros of system (5.1) [32, 86].

By invariant zeros of the system

$$\begin{cases} \dot{x} = Ax + Df \\ y = Cx \end{cases} \tag{5.9}$$

we call the values $s \in \mathbb{C}$ which lower the rank of the Rosenbrock system matrix $R(s)$, i.e., such values that

$$\operatorname{rank} R(s) = \operatorname{rank} \begin{pmatrix} sI - A & -D \\ C & 0 \end{pmatrix} < n + m,$$

where $R(s) \in \mathbb{C}^{(n+l)\times(n+m)}$.

Invariant zeros define zero dynamics of system (5.9), i.e., its dynamics under the condition $y(t) \equiv 0$.

In the case where $l > m$, we use the following algorithm for determining the characteristic polynomial of zero dynamics. Let us consider various square "subsystems" (5.9), i.e., systems of the form

$$\begin{cases} \dot{x} = Ax + Df \\ y_{i_1\dots i_m} = C_{i_1\dots i_m}x, \quad i_1,\dots,i_m \in \{1,\dots,l\}, \quad i_p \neq i_q \text{ for } p \neq q, \end{cases}$$

where $C_{i_1\dots i_m} = \begin{pmatrix} C_{i_1} \\ \vdots \\ C_{im} \end{pmatrix} \in \mathbb{R}^{m\times n}$ is a matrix formed by the rows $C_{i_1}, \dots, C_{i_m}$ of the matrix C, $y_{i_1\dots i_m} = \begin{pmatrix} y_{i_1} \\ \vdots \\ y_{im} \end{pmatrix} \in \mathbb{R}^m$ being an m-dimensional output, a part of the full vector $y(t) \in \mathbb{R}^l$. Each of systems indicated above, corresponding to $i_1, \dots, i_m$, is square, i.e., the dimension of the output $y_{i_1\dots i_m}$ for it coincides with that of the input f. Therefore, for each one of the indicated systems the characteristic polynomial of zero dynamics $\beta_{i_1,\dots,i_m}(s)$ is a determinant of the corresponding Rosenbrock matrix

$$\beta_{i_1\dots i_m}(s) = \det\begin{pmatrix} sI - A & -D \\ C_{i_1\dots i_m} & 0 \end{pmatrix}.$$

The characteristic polynomial $\beta(s)$ of zero dynamics of system (5.9) with respect to the full l-dimensional output $y(t)$ is the largest common divider of all polynomials $\beta_{i_1\dots i_m}(s)$ (they are C_l^m in number). In the general case, for $l > m$, the polynomial $\beta(s) = 1$, i.e., the system does not have invariant zeros (to be more precise, among all systems (5.9) systems with invariant zeros form a set of measure zero). If invariant zeros of a system are absent or stable (i.e., the zero dynamics of the system is absent or stable, respectively), then system (5.9) is said to be a *minimal-phase* system.

In the sequel we shall assume that the following assumption is fulfilled.

Assumption A.5. System (5.9) is minimal-phase, i.e., its invariant zeros are absent or lie in $\mathbb{C}_-$.

The following statement holds for system (5.9).

Theorem 5.1. *Suppose that Assumptions* A.1–A.4 *are fulfilled for system* (5.9). *Then, if the system does not have invariant zeros, the pair* $\{C_1'', A_{11}\}$ *is observable. If the system has invariant zeros, then they form an unchangeable part of the spectrum of the matrix* $A_L = A_{11} - LC_1''$; *if the invariant zeros are stable, then the pair* $\{C_1'', A_{11}\}$ *is reconstructible.*

Proof. Let us consider in greater detail the Rosenbrock matrix of system (5.9). Note that the set of invariant zeros of the system is invariant relative to the nondegenerate change of coordinates and nonsingular transformations of the input and output. Therefore, in order to determine the set of invariant zeros it suffices to write the Rosenbrock matrix for the system reduced to form (5.4). In this case

$$y = \begin{pmatrix} y' \\ y'' \end{pmatrix} = \begin{pmatrix} 0 & I_m \\ C_1'' & C_2'' \end{pmatrix} x$$

and we can write the Rosenbrock matrix in the block form

$$R(s) = \begin{pmatrix} sI_{n-m} - A_{11} & A_{12} & 0 \\ -A_{21} & sI_m - A_{22} & -(C'D) \\ 0 & I_m & 0 \\ C_1'' & C_2'' & 0 \end{pmatrix}.$$

According to Assumption A.4 $\det(C'D) \neq 0$, and therefore $\operatorname{rank} R(s) = n + m$ if and only if

$$\operatorname{rank} \begin{pmatrix} sI_{n-m} - A_{11} & A_{12} \\ 0 & I_m \\ C_1'' & C_2'' \end{pmatrix} = \operatorname{rank} R'(s) = n.$$

Let us perform transformations of the matrix $R'(s)$ which do not change its rank, namely, let us subtract the second block row multiplied by $(-A_{12})$ and $(-C_2'')$ from the first and third block rows, respectively. Then we obtain

$$\operatorname{rank} R'(s) = \operatorname{rank} \begin{pmatrix} sI_{n-m} - A_{11} & 0 \\ 0 & I_m \\ C_1'' & 0 \end{pmatrix} = \operatorname{rank} R''(s).$$

It is obvious that $\operatorname{rank} R''(s) = n$ if and only if

$$\operatorname{rank} \begin{pmatrix} sI_{n-m} - A_{11} \\ C_1'' \end{pmatrix} = \operatorname{rank} R'''(s) = n - m,$$

i.e., the matrix $R'''(s)$ is of full rank. The fullness of the rank of the matrix $R'''(s)$ for all $s \in \mathbb{C}$ under the condition of the fullness of the rank of C_1'' (and this condition is fulfilled) is a necessary and sufficient condition of the observability of the pair $\{C_1'', A_{11}\}$. The points of lowering the rank of the matrix $R'''(s)$ define the spectrum of the nonobservable dynamics of the pair $\{C_1'', A_{11}\}$.

It follows from the nonsingular transformations performed above that the rank of the Rosenbrock matrix $R(s)$ is lost if and only if the rank of the matrix $R'''(s)$ is lost, the sets of points of lowering the rank of these matrices being coincident. Thus, the zero dynamics of the original system defines the nonobservable dynamics of the pair $\{C_1'', A_{11}\}$. The proof of the theorem is complete. □

Remark 5.2. We can obtain the same result using representation (5.4). This approach allows us to write explicitly the structure of the observer.

Let us consider in greater detail the fully determined system (5.7). Note that since the pair $\{A, D\}$ is controllable, the pair $\{A_{11}, A_{12}\}$ in system (5.7) must also be controllable (i.e., the first subsystem in system (5.4) is controllable by means of $y'(t)$).

Then there exists a Kalman decomposition for system (5.7) which divides this system into an observable and nonobservable parts

$$\begin{cases} \dot{x}_1' = A_{11}' x_1' + A_{12}' y' \\ \dot{x}_2' = A_{11}'' x_1' + A_{11}''' x_2' + A_{12}'' y', \end{cases} \tag{5.7*}$$

where x_1' is the observable part of system (5.7) and x_2' is the nonobservable part of the system (if it exists, i.e., if the pair $\{C_1'', A_{11}\}$ is nonobservable). In this case (with an accuracy to within a nondegenerate change of coordinates)

$$x' = \begin{pmatrix} x_1' \\ x_2' \end{pmatrix}, \quad A_{11} = \begin{pmatrix} A_{11}' & 0 \\ A_{11}'' & A_{11}''' \end{pmatrix}, \quad A_{12} = \begin{pmatrix} A_{12}' \\ A_{12}'' \end{pmatrix}, \quad C_1'' = (\tilde{C}_1''; 0).$$

Since the remaining part of the phase vector of the full system (5.4) is a part of the output x', it is obvious that the zero dynamics of the full system (5.4) coincides with the zero dynamics of system (5.7*) (for the input $y' \equiv 0$). Let us investigate it.

For such a motion $y' \equiv 0$, $\tilde{y}'' \equiv 0$ and, in addition, since the part x_1' from $\tilde{y}'' \equiv 0$ is observable, it follows that $x_1' \equiv 0$ for $t \geq 0$. Therefore the zero dynamics of system (5.7*) for $y' \equiv 0$ (and, consequently, the zero dynamics of the original system (5.4)) is defined by the nonobservable part of system (5.7*) and is described by the equation

$$\dot{x}_2' = A_{11}''' x_2'.$$

Thus, if the zero dynamics of the system is absent, then system (5.7*) does not have a nonobservable part and the pair $\{C_1'', A_{11}\}$ is observable. Now if the original system has stable invariant zeros (or, what is the same, stable zero dynamics), then the nonobservable part of system (5.7*) is stable and the pair $\{C_1'', A_{11}\}$ is reconstructible.

5.2 Functional observers

The approach proposed above allows us to solve the problem of synthesis of a functional observer for system (5.9) as well in the case where the number of outputs exceeds the number of unknown inputs, i.e., $l > m$.

Consider a problem of reconstructing the functional $\sigma = Fx$, $\sigma \in \mathbb{R}^p$, where $F \in \mathbb{R}^{p \times n}$ is a known matrix. Let, as before, Assumptions A.1–A.4 be fulfilled for system (5.9). Then, as was shown above, by a nonsingular transformation the system is reduced to form (5.4). After the indicated change of coordinates with matrix P the functional $\sigma = Fx$ assumes the form

$$\sigma = Fx = FP \begin{pmatrix} x' \\ y' \end{pmatrix} = F'x' + F''y' = \sigma' + \sigma'',$$

where $F' \in \mathbb{R}^{p \times (n-m)}$, $F'' \in \mathbb{R}^{p \times m}$ are known matrices. Note that the functional $\sigma'' = F''y'$ is known, and, consequently, in order to construct an estimate for the functional σ it suffices to construct an estimate for its unknown part $\sigma' = F'x'$.

For this purpose it is sufficient to consider again the reduced system without uncertainty (5.7) of order $(n-m)$:

$$\begin{cases} \dot{x}' = A_{11}x' + A_{12}y' \\ \tilde{y} = C_1''x' \end{cases}$$

for which we have to construct a functional observer for the functional $\sigma' = F'x'$. The methods for solving this problem, including the conditions for constructing a minimal order functional observer were given in detail in Chap. 4. The conditions of observability for this system are given by Theorem 5.1.

5.3 Synthesis of observers by the method of pseudoinputs

Under the conditions imposed on system (5.9) indicated above for synthesizing an observer of the full-phase vector of system $x(t)$ we can use the approach based on the decomposition of the system with the employment of the so-called pseudoinputs. We shall describe this method in detail following [13].

Suppose that, as before, Assumptions A.1–A.4 are fulfilled for the system and the invariant zeros of the system are absent or stable (i.e., condition A.5 is fulfilled). Since the number l of outputs of the system exceeds the number of the unknown inputs m, we complement the system $(l-m)$ by "pseudoinputs" (virtual inputs) $f' \in \mathbb{R}^{l-m}$. As a result we obtain a square system with l inputs and l outputs

$$\begin{cases} \dot{x} = Ax + Df + D'f' = Ax + \bar{D}\bar{f} \\ y = Cx, \end{cases} \tag{5.10}$$

where $\bar{f} = \begin{pmatrix} f \\ f' \end{pmatrix}$, $\bar{D} = (D \;\; D')$, the technique of choosing the matrix $D' \in \mathbb{R}^{n\times(l-m)}$ will be described in the sequel.

Note that for $f' \equiv 0$ system (5.10) coincides with system (5.9), and therefore the observer constructed for system (5.10) under the condition $f' \equiv 0$ will reconstruct the phase vector of the original system.

Since system (5.10) is square, the characteristic polynomial of its zero dynamics is a determinant of the Rosenbrock matrix

$$\beta(s) = \det R'(s) = \det \begin{pmatrix} sI_n - A & -(D \;\; D') \\ C & 0 \end{pmatrix}.$$

If the polynomial $\beta(s)$ which depends on the choice of the matrix D' is of order $(n-l)$ (i.e., $\det(C\bar{D}) \neq 0$), then for system (5.10) we can carry out the decomposition with matrix P described above with the isolation of zero dynamics, and then the system will assume the form

$$\begin{cases} \dot{x}' = A_{11}x' + A_{12}y \\ \dot{y} = A_{21}x' + A_{22}y + C\bar{D}\bar{f}, \end{cases}$$

where $x' \in \mathbb{R}^{n-l}$ and $\det(sI - A_{11}) = \beta(s)$. If the matrix D' is chosen such that the polynomial $\beta(s)$ is a Hurwitz polynomial, then the problem of reconstruction of the unknown part of the phase vector x' is solved by an observer of order $(n-l)$:

$$\dot{\tilde{x}}' = A_{11}\tilde{x}' + A_{12}y. \tag{5.11}$$

In this case, the observation error $e' = \tilde{x}' - x'$ satisfies the equation

$$\dot{e}' = A_{11}e',$$

and, consequently, the rate of convergence of the observer is wholly defined by the degree of stability of the polynomial $\beta(s)$. The dimension of observer (5.11) coincides with that of the minimal-order observer described above (where a Luenberger observer is constructed for the fully determined reduced system (5.7)).

Thus, the problem reduces to the search for a matrix D' such that the polynomial $\beta(s)$ will be of degree $(n-l)$ and be a Hurwitz polynomial. Then the following statement holds.

Theorem 5.3. *Suppose that Assumptions* A.1–A.5 *are fulfilled for system* (5.9). *Then, if the system does not have invariant zeros, the roots of the polynomial $\beta(s)$ of order $(n-l)$ are fully determined by the choice of the matrix D'. If the system has invariant zeros, then they are roots of the polynomial $\beta(s)$, the other roots being defined by the choice of the matrix D'.*

Remark 5.4. Thus, the necessary condition for $\beta(s)$ to be a Hurwitz polynomial is the fact that system (5.9) is of minimal phase.

Proof. Consider a Rosenbrock matrix of the extended system (5.10):

$$R'(s) = \begin{pmatrix} sI - A & -D & -D' \\ C & 0 & 0 \end{pmatrix} = \left(\bar{R}'(s) \,\middle|\, \begin{matrix} -D' \\ 0 \end{matrix} \right),$$

where $\bar{R}'(s)$ is a Rosenbrock matrix of system (5.9). Note that if the point s^* is an invariant zero of system (5.9) (i.e., rank $\bar{R}'(s^*) < n+m$), then rank $R'(s^*) < n+l$ as well, i.e., s^* is a root of the polynomial $\beta(s) = \det R(s)$. Consequently, all invariant zeros of system (5.9) are contained in the set of roots of the polynomial $\beta(s)$ for any choice of D'.

In what follows, without loss of generality, we shall assume that the full-rank matrix D has the form

$$D = \begin{pmatrix} 0 \\ I_m \end{pmatrix}, \quad I_m \in \mathbb{R}^{m\times m}.$$

This can always be achieved by means of a nondegenerate change of coordinates and inputs of the system. Let us represent the matrices of system (5.9) in block form

$$A = \begin{pmatrix} A_1 & A_2 \\ A_3 & A_4 \end{pmatrix} \begin{matrix} \}n-m \\ \}m \end{matrix}, \quad D' = \begin{pmatrix} D_1' \\ D_2' \end{pmatrix} \begin{matrix} \}n-m \\ \}m \end{matrix}, \quad C = (\underbrace{C_1}_{n-m} \mid \underbrace{C_2}_{m})$$

and write the Rosenbrock matrix $R'(s)$ also in block form

$$R'(s) = \begin{pmatrix} sI_{n-m} - A_1 & -A_2 & 0 & -D_1' \\ -A_3 & sI_m - A_4 & -I_m & -D_2' \\ C_1 & C_2 & 0 & 0 \end{pmatrix} \begin{matrix} \}n-m \\ \}m \\ \}l \end{matrix} .$$
$$\qquad\qquad \underbrace{\quad}_{n-m} \quad \underbrace{\quad}_{m} \quad \underbrace{\quad}_{m} \quad \underbrace{\quad}_{l-m}$$

Then it is obvious that rank $R' = m + \operatorname{rank} R''(s)$, where

$$R''(s) = \begin{pmatrix} sI_{n-m} - A_1 & -A_2 & -D_1' \\ C_1 & C_2 & 0 \end{pmatrix} = \left(\bar{R}''(s) \middle| \begin{matrix} -D_1' \\ 0 \end{matrix} \right).$$

Note that rank $\bar{R}'(s) = m + \operatorname{rank} \bar{R}''(s)$ for the original system, and, consequently, the set of points of lowering the rank $\bar{R}''(s)$ defines the set of invariant zeros of the original system.

According to Assumption A.4,

$$\operatorname{rank} CD = \operatorname{rank} \left((C_1 \;\; C_2) \begin{pmatrix} 0 \\ I_m \end{pmatrix} \right) = \operatorname{rank} C_2 = m.$$

Since $C_2 \in \mathbb{R}^{l \times m}$ and $l > m$, we can use the nondegenerate transformation of outputs (i.e., rows of the matrix C) to reduce the matrices C_1 and C_2 to the form

$$C_1 = \begin{pmatrix} C_1' \\ C_2'' \end{pmatrix}, \quad C_2 = \begin{pmatrix} I_m \\ 0 \end{pmatrix}, \quad C_1' \in \mathbb{R}^{m \times (n-m)}, \quad C_1'' \in \mathbb{R}^{(l-m) \times (n-m)}.$$

Then

$$R''(s) = \begin{pmatrix} sI_{n-m} - A_1 & -A_2 & -D_1' \\ C_1' & I_m & 0 \\ C_1'' & 0 & 0 \end{pmatrix} \begin{matrix} \}n-m \\ \}m \\ \}l-m \end{matrix}$$
$$\qquad\qquad \underbrace{\quad}_{n-m} \quad \underbrace{\quad}_{m} \quad \underbrace{\quad}_{l-m}$$

and

$$\operatorname{rank} R''(s) = m + \operatorname{rank} R'''(s),$$

where

$$R'''(s) = \begin{pmatrix} sI_{n-m} - \tilde{A} & -D_1' \\ C_1'' & 0 \end{pmatrix} = \left(\bar{R}'''(s) \middle| \begin{matrix} -D_1' \\ 0 \end{matrix} \right), \quad \tilde{A} = A_1 - A_2 C_1'.$$

In this case, the set of points of lowering the rank of the matrix $\bar{R}'(s)$ (and $\bar{R}''(s)$) coincides with the corresponding set of points of the matrix $\bar{R}'''(s)$.

Consider an $(n-m)$-dimensional linear stationary system with $(l-m)$ inputs and $(l-m)$ outputs:

$$\begin{cases} \dot{z} = \tilde{A}z + D_1'\omega \\ e = C_1''z, \end{cases} \tag{5.12}$$

for which the matrix $R'''(s)$ is a Rosenbrock matrix. On the other hand, the matrix $\bar{R}'''(s)$ defines the observability (reconstructibility) of the pair $\{C_1'', \tilde{A}\}$. Consequently, the pair $\{C_1'', \tilde{A}\}$ is observable if and only if the original system does not have invariant zeros (when the matrices $\bar{R}'(s)$ and $\bar{R}'''(s)$ are of full rank for all $s \in \mathbb{C}$). The pair $\{C_1'', \tilde{A}\}$ is reconstructible if and only if the invariant zeros of the matrix $\bar{R}'(s)$ are stable, the set of invariant zeros of the original system being coincident with the spectrum of the nonobservable part of system (5.12).

In addition, since the transformations carried out are nondegenerate, the sets of points of degeneracy of the Rosenbrock matrix $R''(s)$ of the system with pseudoinputs (5.10) and the matrix $R'''(s)$ coincide, moreover, the relation

$$\beta(s) = \det R'(s) = \pm \det R'''(s)$$

holds with an accuracy to within the sign, i.e., the characteristic polynomials of the zero dynamics of systems (5.10) and (5.12) coincide.

Let us consider the zero dynamics of system (5.12) in greater detail. Since the matrix C is of full rank, the matrix $C_1'' \in \mathbb{R}^{(l-m)\times(n-m)}$ is also of full rank, i.e., rank $C_1'' = l - m$. Let us carry out in system (5.12) a nondegenerate change of coordinates with matrix Q:

$$\begin{pmatrix} z' \\ e \end{pmatrix} = Qz, \quad z = Q^{-1}\begin{pmatrix} z' \\ e \end{pmatrix}, \quad z' \in \mathbb{R}^{n-l}.$$

In the new coordinates system (5.12) assumes the form

$$\begin{cases} \dot{z}' = \tilde{A}_{11}z' + \tilde{A}_{12}e + D_{11}'\omega \\ \dot{e} = \tilde{A}_{21}z' + \tilde{A}_{22}e + D_{12}'\omega, \end{cases} \tag{5.13}$$

where the matrices D_{11}' and D_{12}' are defined by the choice of D_{11}':

$$\begin{pmatrix} D_{11}' \\ D_{12}' \end{pmatrix} = D_1'Q, \quad D_1' = \begin{pmatrix} D_{11}' \\ D_{12}' \end{pmatrix} Q^{-1}.$$

Since the matrix Q is nondegenerate, the inverse statement, namely, that the matrix D_1' is wholly defined by the choice of D_{11}' and D_{12}', also holds true.

Let us choose the matrix $D_{12}' \in \mathbb{R}^{(n-m)\times(n-m)}$ nondegenerate, say, $D_{12}' = I_{n-m}$. We find the equation of zero dynamics of system (5.13) under this condition. Since $\dot{e} = e \equiv 0$ in this case, the second equation (5.13) gives a relation $\omega = -\tilde{A}_{21}z'$, and, consequently, the zero dynamics is defined by the equation

$$\dot{z}' = (\tilde{A}_{11} - D_{11}'\tilde{A}_{21})z'.$$

On the other hand, the Rosenbrock matrix of system (5.13) has block structure:

$$R'''' = \begin{pmatrix} sI_{n-l} - \tilde{A}_{11} & -\tilde{A}_{12} & -D_{11}' \\ -\tilde{A}_{21} & sI_{l-m} - \tilde{A}_{22} & -D_{12}' \\ 0 & I_{l-m} & 0 \end{pmatrix}.$$

Taking into account that system (5.13) is the transformed system (5.12), we obtain an identity $\det R'''(s) = \det R''''(s)$. In the matrix $R''''(s)$ we shall consider a submatrix

$$\bar{R}''''(s) = \begin{pmatrix} sI_{n-l} - \tilde{A}_{11} & -\tilde{A}_{12} \\ -\tilde{A}_{21} & sI_{l-m} - \tilde{A}_{22} \\ 0 & I_{l-m} \end{pmatrix}.$$

The set of degeneration points of the rank of the matrix $\bar{R}''''(s)$ coincides with the corresponding set of the matrix $\bar{R}'''(s)$ on one hand (and, consequently, the matrix $\bar{R}'(s)$), and, on the other hand, is defined by the degeneration of the matrix

$$\bar{R}^V(s) = \begin{pmatrix} sI_{n-l} - \tilde{A}_{11} \\ -\tilde{A}_{21} \end{pmatrix}.$$

The indicated matrix is connected with the observability of the pair $\{\tilde{A}_{21}, \tilde{A}_{11}\}$, namely, if the original system does not have invariant zero, then the pair $\{\tilde{A}_{21}, \tilde{A}_{11}\}$ is observable and if the original system has stable invariant zeros, then the pair $\{\tilde{A}_{21}, \tilde{A}_{11}\}$ is reconstructible.

To complete the proof of the theorem, it remains to note that the matrix of zero dynamics of system (5.13) has the form

$$A_{\text{zd}} = \tilde{A}_{11} - D'_{11}\tilde{A}_{21},$$

and, consequently, in the case where the original system does not have invariant zeros, the spectrum of the matrix A_{zd} can be defined arbitrarily by the choice of D'_{11} (since $\{\tilde{A}_{21}, \tilde{A}_{11}\}$ is observable). Now if the original system has stable invariant zeros, then, by virtue of reconstructibility of the pair $\{\tilde{A}_{21}, \tilde{A}_{11}\}$, the spectrum of A_{zd} can be made stable by the choice of D'_{11} (in this case, a part of the spectrum coincides with the invariant zeros of the original system and the remaining part can be defined arbitrarily). The theorem is proved. □

Remark 5.5. Under the pseudodisturbance of $f' \equiv 0$ the statement of Theorem 5.3 can be formulated in an equivalent form.

Theorem 5.3′. *If Assumptions* A.1–A.4 *are fulfilled for system* (5.9) *and the system is minimal-phase, then, by means of a nonsingular transformation, it can be reduced to the form*

$$\begin{cases} \dot{x}' = A_{11}x' + A_{12}y \\ \dot{y} = A_{21}x' + A_{22}y + CDf, \end{cases} \tag{5.14}$$

where the spectrum of the matrix A_{11} contains all invariant zeros of system (5.9) *and its remaining part can be chosen arbitrarily (and is defined by the choice of the transformation matrix).*

We should also note that the number of pseudoinputs can be any number ρ from 1 to $l-m$. In this case, we define, according to the output, the $(m+\rho)$-dimensional output such that according to it properties A.1–A.4 are fulfilled for this output. For this output we can use the algorithm of decomposition of the system proposed above.

The following statement is valid.

Theorem 5.3″. *Suppose that Assumptions* A.1–A.4 *are fulfilled for system* (5.9), $1 \le \rho \le l-m$, *and system* (5.9) *is minimal-phase relative to the output* $y_\rho = C_\rho x \in \mathbb{R}^{m+\rho}$, *where* C_ρ *is a matrix formed by the rows of the matrix* C. *Then, by means of nonsingular transformation the system can be reduced to the form*

$$\begin{cases} \dot{x}_\rho = A_{11}x_\rho + A_{12}y_\rho \\ \dot{y}_\rho = A_{21}x_\rho + A_{22}y_\rho + C_\rho Df, \end{cases}$$

where A_{11} *is a Hurwitz matrix,* $x_\rho \in \mathbb{R}^{n-m-\rho}$.

5.4 Classical methods of synthesis of observers under the uncertainty conditions

Many different methods have been proposed for solving the problem of synthesis of an observer for an uncertain system (5.1) under the condition $l > m$. However, practically all of them give a solution under the same conditions imposed on system (5.1). These conditions are indicated in Theorems 5.1 and 5.3. We shall give a brief review of these methods without detailed proofs. The results will be given only for the case $u \equiv 0$, i.e., for system (5.9).

The analysis of published works shows that with a certain conventionality all known methods of synthesis of observers for systems of this kind use one of the following ideas:

(1) removal of disturbance from the equation of error estimation,

(2) an isolation of a subsystem from the system which does not explicitly depend on the disturbance,

(3) the use of special canonical forms of multiconnected systems,

(4) the solution of the observation problem as a problem of stabilization,

(5) the use of fictitious (or pseudo-) inputs (outputs).

We shall consider each of these methods and give the main results.

5.4.1 Removal of disturbance from the equation of estimation error

Observers of this kind are proposed in [115, 41] and by their structure are similar to the structure of classical Luenberger observers for systems without uncertainty. For solving the observation problem for system (5.9) the authors propose to use an observer of the form

$$\begin{cases} \dot{z} = Ez + Fy \\ \tilde{x} = Hz + Ly, \end{cases} \tag{5.15}$$

where $z \in \mathbb{R}^p$, $\tilde{x} \in \mathbb{R}^n$, H, E, F, and L are constant matrices of the requisite dimensions which must be determined. For a full-dimensional observer the dimension of the vector z is equal to that of x, i.e., $p = n$, and we can set $H = I$. In this case, as an estimate of the unknown vector $x(t)$ we can use the output of the observer $\tilde{x}(t)$. Then the observation error $\mathcal{E}(t) = x(t) - \tilde{x}(t)$ satisfies the equation

$$\dot{\mathcal{E}} = E\mathcal{E} + (PA - FC - EP)x + PDf,$$

where $P = I - LC$. The matrices of the observer are chosen such that we can exclude from the last equation the unknown disturbance (signal f) and the unknown phase vector $x(t)$. This condition will be fulfilled if the relations

$$PA - FC - EP = 0, \quad PD = 0 \tag{5.16}$$

hold. In this case, the error satisfies the equation

$$\dot{\mathcal{E}} = E\mathcal{E},$$

and, consequently, $\mathcal{E}(t)$ tends to zero exponentially under the condition that E is a Hurwitz matrix. The following statement is proved in [41].

Statement 5.6. If Assumptions A.1–A.4 are fulfilled for system (5.9), then there exist matrices E, F, and L which satisfy conditions (5.16). Moreover, if Assumption A.5 is fulfilled, then the matrix E can be chosen as a Hurwitz matrix. A part of its spectrum coincides with the invariant zeros of the Rosenbrock matrix of the original system and the remaining part is chosen arbitrarily.

The drawback of the proposed approach is that we have to solve a cumbersome system of matrix equations (5.16) whose solution is, generally speaking, not unique. In addition, the dimension of the observer constructed in [41] is equal to the dimension of the system n, it exceed the dimension of the observer of the lowered order $(n - l)$ which can be achieved by other methods. Obviously, this approach can be generalized to the case $p < n$, in particular, $p = n - l$. Moreover, it can also be generalized to the problem of construction of functional observers. This problem was considered in [104] where the functional $\sigma = Fx$, $F \in \mathbb{R}^{q \times n}$ was reconstructed. Here is the main idea of this work.

Suppose that we know the decomposition of the matrix F of the form

$$F = KT + WC,$$

where $K \in \mathbb{R}^{q\times p}$, $T \in \mathbb{R}^{p\times n}$, $W \in \mathbb{R}^{q\times l}$, and $q \le p \le n$. We denote $x' = Tx \in \mathbb{R}^p$, and then

$$\sigma(t) = Fx(t) = Kx' + Wy.$$

In order to reconstruct the functional $\sigma(t)$ it suffices to construct an estimate x' which is given by the observer of order p similar in structure to observer (5.15):

$$\begin{cases} \dot{z} = Ez + Gy \\ \tilde{\sigma} = Kz + Wy, \end{cases} \tag{5.17}$$

where the matrices $E \in \mathbb{R}^{p\times p}$ and $G \in \mathbb{R}^{p\times l}$ must be defined and $z(t)$ is the estimate of the vector $x'(t)$. The estimation error $e(t) = z(t) - x'(t)$ satisfies the equation

$$\dot{e} = Ee + (GC - TA + ET)x - TDf$$

and the parameters can be found from the relations

$$\begin{cases} GC - TA + ET = 0 \\ TD = 0 \\ E \ \text{ is a Hurwitz matrix.} \end{cases} \tag{5.18}$$

The following statement is proved in [104].

Statement 5.7. Suppose that Assumptions A.1–A.5 are fulfilled for system (5.9). Then there exists a solution of system (5.18) and observer (5.17) gives an exponential estimate of the unknown functional $\sigma(t)$.

The authors of [104] proposed a procedure for solving system (5.18) for the stable matrix E and proved the existence of a solution of this system, but only under the condition that

$$p \ge \frac{q(n-l)}{l-m}.$$

Note that for $q > l - m$ the inequality $p > n - l$ is satisfied, i.e., the dimensional of the functional observer exceeds that of the observer for a full-phase vector. This is obviously connected with the method of proving the statement and not with the properties of the method.

5.4.2 The method of removal of disturbance from the equation of the system

Another approach to the construction of asymptotic observers is connected with the transformation of the system which removes the explicit presence of disturbance in the equations of the system. This method is realized in [51, 54, 55].

The main idea of the method is the transformations of equations of the system so that the explicit presence of $f(t)$ (or a function of it) which obviously given information about the signal $f(t)$ is replaced by the known signal $y(t)$.

Let us consider again the original system (5.9). If Assumptions A.1–A.4 are fulfilled in this system, we introduce a new variable

$$\xi = x - Hy.$$

Then the equation for ξ can be written as

$$\dot{\xi} = (A - HCA)\xi + (A - HCA)Hy + (D - HCD)f$$

with an equation for the output

$$y = C\xi + CHy \quad \text{or} \quad (I - CH)y = C\xi.$$

If the matrix H is such that $(D - HCD) = 0$ and $CH - I \neq 0$, then the equations for the new variables ξ do not explicitly depend on f:

$$\begin{cases} \dot{\xi} = \bar{A}\xi + \bar{A}Hy, & \bar{A} = A - HCA \\ \bar{y} = C\xi, & \bar{y} = (I - CH)y, \end{cases} \tag{5.19}$$

and the original problem reduces to the observation problem for a system with the known input $y(t)$ which can be solved by standard methods if the pair $\{C, \bar{A}\}$ is observable (reconstructible).

The following statement is proved in [54].

Statement 5.8. Suppose that Assumptions A.1–A.5 are fulfilled for system (5.9). Then, if system (5.9) does not have invariant zeros, then the pair $\{C, \bar{A}\}$ is observable and if this system has invariant zeros in $\mathbb{C}_-$, then the pair $\{C, \bar{A}\}$ is reconstructible, and the unchangeable eigenvalues of the matrix $A_L = \bar{A} - LC$ coincide with the invariant zeros of the original system.

For system (5.19) we can synthesize standard observers of full order n as well as of a lowered order $(n - l)$. We take $\tilde{x} = \tilde{\xi} + Hy$, where $\tilde{\xi}$ is an estimate for ξ, as an estimate of x.

5.4.3 Methods based on the reduction of a system to a special canonical form

We shall give algorithms for synthesizing observers based on the reduction of a system to a special form where a part of variables do not depend explicitly on the disturbance $f(t)$ and can be reconstructed.

We know several approaches to this transformation. One of them was proposed in [48]. Here is this method.

We begin with transforming system (5.9) using the matrix $T \in \mathbb{R}^{n\times n}$ such that

$$TD = \begin{pmatrix} T_1 \\ T_2 \end{pmatrix} D = \begin{pmatrix} 0 \\ \bar{D} \end{pmatrix},$$

where $T_1 \in \mathbb{R}^{(n-m)\times n}$, $T_2 \in \mathbb{R}^{m\times n}$, $\bar{D} \in \mathbb{R}^{m\times m}$, $\det \bar{D} \neq 0$. Since the matrix D is of full rank, such a matrix T always exists. The transformation $\binom{x'}{x''} = Tx$ reduces system (5.9) to the form

$$\begin{cases} \begin{cases} x' = A_1 x' + A_2 x'' \\ x'' = A_3 x' + A_4 x'' + \bar{D} f \end{cases} \\ \quad y = C_1 x' + C_2 x''. \end{cases} \tag{5.20}$$

Note that if Assumption A.4 is fulfilled, then $\operatorname{rank} C_2 = m$. In addition, $C_2 \in \mathbb{R}^{l\times m}$, $C_1 \in \mathbb{R}^{l\times(n-m)}$, and the equation for x' does not explicitly depend on the disturbance $f(t)$.

By virtue of the remarks that we have made, there exist matrices $Q \in \mathbb{R}^{l\times l}$ and $P \in \mathbb{R}^{m\times m}$ such that

$$QC_2P = \begin{pmatrix} Q_1 \\ Q_2 \end{pmatrix} C_2 P = \begin{pmatrix} \bar{C}_2 \\ 0 \end{pmatrix}, \quad \det \bar{C}_2 \neq 0, \quad \det P \neq 0.$$

The matrix Q performs the transformation of outputs y and the matrix P of parts of the phase variables x''. The authors of [48] proposed a procedure of construction of matrices P and Q. Under the assumptions that we have made, the output y is reduced to the form

$$\begin{pmatrix} y_1 \\ y_2 \end{pmatrix} = Qy = \begin{pmatrix} Q_1 y \\ Q_2 y \end{pmatrix} = \begin{pmatrix} Q_1 C_1 x' + \bar{C}_2 P^{-1} x'' \\ Q_2 C_1 x' \end{pmatrix},$$

and, consequently, a part of the phase vector x'' can be expressed via the vector x' and the component of the output $y_1 = Q_1 y$, i.e.,

$$x'' = P\bar{C}_2^{-1}(y_1 - Q_1 C_1 x').$$

Substituting the expression for x'' into the first equation (5.20), we obtain a system for x' of order $(n-m)$

$$\begin{aligned} \dot{x}' &= A_1 x' + A_2 P \bar{C}_2^{-1}(y_1 - Q_1 C_1 x') \\ &= \underbrace{\left(A_1 - A_2 P \bar{C}_2^{-1} Q_1 C_1\right)}_{\tilde{A}} x' + A_2 P \bar{C}_2^{-1} y_1 \end{aligned} \tag{5.21}$$

with the known input $y_1(t)$ which does not depend explicitly on the disturbance $f(t)$ and output $y_2(t)$ of order $(l-m)$, i.e.,

$$y_2 = Q_2 C_1 x' = \tilde{C} x'. \tag{5.22}$$

The following statement holds true.

Statement 5.9. Suppose that Assumptions A.1–A.5 are fulfilled for system (5.9). Then the pair $\{\tilde{C}, \tilde{A}\}$ is observable if the original system does not have invariant zeros and the pair $\{\tilde{C}, \tilde{A}\}$ is reconstructible if the original system has invariant zeros in $\mathbb{C}_-$ and these zeros belong to the spectrum of the matrix $A_L = \tilde{A} - L\tilde{C}$.

For system (5.21), (5.22) we can construct both an observer of full order $(n - m)$ and that of a lowered order $(n - m) - (l - m) = n - l$. These observers give an estimate $\tilde{x}'$ of a part of the phase vector x' and the relation

$$\tilde{x}'' = P\bar{C}_2^{-1}(y_1 - Q_1 C_1 \tilde{x}').$$

gives an estimate for x''.

The method of quasisplittings described above in detail is also based on the idea of reduction of a system to a special canonical form. Note that it is simpler in realization and, in addition, can be used for solving a problem of construction of functional observers, those of minimal order inclusive.

5.4.4 Method of pseudoinputs

This method was described above in detail, it also makes it possible to reduce a system to a special representation. This approach is due to S. K. Korovin and is described in [13].

5.4.5 Methods of synthesis of observers with control

One more approach to the construction of observers for uncertain systems is based on the use in the observer of control which carries out stabilization at the zero of the system in deviations which certainly depends on the disturbance f. Different methods of stabilization are used for uncertain systems.

For the first time this approach was obviously used in [109, 100]. We shall describe it in greater detail.

For reconstruction of the phase vector of system (5.9) we shall use an observer

$$\dot{\tilde{x}} = A\tilde{x} + L(y - C\tilde{x}) - v, \tag{5.23}$$

where v is a control carrying out the stabilization of the system in deviations $e = \tilde{x} - x$ defined by the equation

$$\dot{e} = (A - LC)e - v - Df = A_L e - (v + Df).$$

The matrix L is chosen from the condition that A_L is a Hurwitz matrix (this choice is always possible since the pair $\{C, A\}$ is observable).

The authors of [109] introduce an additional assumption concerning the unknown disturbance.

Assumption A.6. The signal $f(t)$ is uniformly bounded and its majorant, the constant $\rho > 0$, is known (i.e., $|f(t)| \le \rho$ for $t \ge 0$).

In this case, a discontinuous feedback

$$v = \rho D \,\mathrm{Sgn}(W(y - \tilde{y}))$$

is proposed for employment as stabilizing control. Here $W \in \mathbb{R}^{m \times l}$, $\tilde{y} = C\tilde{x}$, and $\mathrm{Sgn}(z) \in \mathbb{R}^m$ is a discontinuous vector-function with components $\mathrm{sgn}(z_i)$. In this case, a system in deviations has the form

$$\dot{e} = A_L e - D(\rho \,\mathrm{Sgn}(WCe) + f). \tag{5.24}$$

The Lyapunov function

$$V(e) = e^\top P e, \quad P > 0,$$

is used for investigation of its stability. By virtue of system (5.24) its derivative has the form

$$\dot{V} = e^\top (PA_L + A_L^\top P)e - 2(\rho \,\mathrm{Sgn}(WCe) + f)D^\top Pe.$$

The following statement is proved in [109].

Statement 5.10. If Assumptions A.1–A.6 are fulfilled for the system (5.9) (and the system does not have invariant zeros), then there exist matrices P, $Q > 0$ as well as parameters of the observer W and L satisfying the system of equations

$$PA_L + A_L^\top P = -Q,$$
$$D^\top P = WC.$$

In this case, $\dot{V} < -\lambda V$, $\lambda = \mathrm{const} > 0$.

The authors of [45] proposed a procedure of constructing the matrices W and L for the observer of form (5.23) which have a more general form of stabilizing control

$$v = \rho G \,\mathrm{Sgn}(Wy - WC\tilde{x})$$

with a varying matrix G instead of the fixed one D in [109].

The drawbacks of this method are additional constraints imposed on the signal $f(t)$.

5.5 Static and unstatic methods of estimation under the conditions of uncertainty

5.5.1 Observers for square systems with uncertainty

The methods of construction of asymptotic observers for systems with uncertainty (5.1) described above essentially used Assumption A.3 stating that the number of outputs l exceeds m which is the dimension of the unknown input $f(t)$. Let us consider now a situation where these dimensions coincide, i.e., $l = m$. Systems of this kind are traditionally called *square* systems since, in this case, the transfer matrix of system (5.1)

$$W(s) = [C(sI - A)^{-1}D] \in \mathbb{C}^{l\times l}$$

from uncertainty $f(t)$ to output $y(t)$ is square.

Scalar systems with the first relative order. We begin with considering a more simple case of a scalar system, i.e., a system with scalar input $f(t)$ and output $y(t)$ (i.e., $l = m = 1$). As before, without loss of generality, we assume that $u(t) \equiv 0$ since the effect produced by the known input can always be compensated in the observer. Thus, we consider a system

$$\begin{cases} \dot{x} = Ax + Df \\ y = Cx, \end{cases} \tag{5.25}$$

where $A \in \mathbb{R}^{n\times n}$, $D \in \mathbb{R}^{n\times 1}$, $C \in \mathbb{R}^{1\times n}$. The transfer function from the input f to the output y

$$W(s) = C(sI - A)^{-1}D = \frac{\beta_m(s)}{\alpha_n(s)} \tag{5.26}$$

is defined for this system. Here $\beta_m(s)$ and $\alpha_n(s)$ are polynomials of s of the corresponding degrees m and n. In this case

$$\alpha_n(s) = \det(sI - A) = s^n + \alpha_n s^{n-1} + \cdots + \alpha_1 \tag{5.27}$$

is a characteristic polynomial of the matrix A, and the polynomial

$$\beta_m(s) = \beta_{m+1}s^m + \beta_m s^{m-1} + \cdots + \beta_1 \tag{5.28}$$

is a characteristic polynomial of the zero dynamics of the system which is a determinant of the Rosenbrock matrix

$$\beta_m(s) = \det\left(\begin{array}{c|c} sI - A & -D \\ \hline C & 0 \end{array}\right).$$

The relative order of system (5.25) is the number $r = n - m$ which is defined by the relations

$$CD = 0,\ CAD = 0,\ \ldots,\ CA^{r-2}D = 0,\ CA^{r-1}D = \beta_{m+1} \neq 0. \tag{5.29}$$

For system (5.25) we assume that its zero dynamics is asymptotically stable, i.e., $\beta_m(s)$ is a Hurwitz polynomial. In this case, the system is of minimal phase. In addition, we shall assume that the pair $\{C, A\}$ is observable and the pair $\{A, D\}$ is controllable, i.e., system (5.25) is in the general position.

In the general case the condition $CD \neq 0$ is fulfilled for system (5.25) (without loss of generality we assume that $CD = 1$ which can always be achieved by normalizing the output), i.e., the relative order of the system $r = 1$, and then $\deg(\beta_m(s)) = m = n-1$. Since the pair $\{A, D\}$ is controllable, system (5.25) can be reduced, by means of nonsingular change of coordinates, to the controllable canonical representation

$$A = \begin{pmatrix} 0 & 1 & 0 & \dots & 0 \\ 0 & 0 & 1 & \dots & 0 \\ \dots & \dots & \dots & \dots & \dots \\ 0 & 0 & 0 & \dots & 1 \\ -\alpha_1 & -\alpha_2 & -\alpha_3 & \dots & -\alpha_n \end{pmatrix}, \quad D = \begin{pmatrix} 0 \\ \vdots \\ 0 \\ 1 \end{pmatrix}, \quad C = (\beta_1, \dots, \beta_n),$$

here $CD = \beta_n = 1$. We carry out a standard change of coordinates with the isolation of zero dynamics of the system for which purpose we pass from the coordinates $(x_1, \dots, x_n)^\top$ to coordinates $(x_1, \dots, x_{n-1}, y)^\top$. Since $\beta_n = 1$, it follows that $y = \beta_1 x_1 + \dots + \beta_{n-1}x_{n-1} + x_n$, and therefore, in the new coordinates, the system assumes the form

$$\begin{cases} \dot{x}_1 = x_2 \\ \dot{x}_2 = x_3 \\ \quad\vdots \\ \dot{x}_{n-2} = x_{n-1} \\ \dot{x}_{n-1} = -\beta_1 x_1 - \beta_2 x_2 - \dots - \beta_{n-1}x_{n-1} + y \\ \dot{y} = -\gamma_1 x_1 - \gamma_2 x_2 - \dots - \gamma_{n-1}x_{n-1} - \gamma_n y + f, \end{cases} \tag{5.30}$$

where $\gamma_n = (\beta_{n-1} - \alpha_n)$, $\gamma_i = -\gamma_n \beta_i + \beta_{i-1} - \alpha_i$, $i = 1, \dots, n-1$. Note that the first $(n-1)$ equations of system (5.30) describe the zero dynamics of the system which does not depend explicitly on the unknown disturbance $f(t)$. In order to reconstruct the first $(n-1)$ coordinates of system (5.30) we use the observer

$$\begin{cases} \dot{\tilde{x}}_1 = \tilde{x}_2 \\ \quad\vdots \\ \dot{\tilde{x}}_{n-2} = \tilde{x}_{n-1} \\ \dot{\tilde{x}}_{n-1} = -\beta_1 \tilde{x}_1 - \beta_2 \tilde{x}_2 - \dots - \beta_{n-1}\tilde{x}_{n-1} + y. \end{cases} \tag{5.31}$$

In this case, the observation error $e' = x' - \tilde{x}'$ (where $x' = (x_1, \dots, x_{n-1})^\top$, $\tilde{x}' = (\tilde{x}_1, \dots, \tilde{x}_{n-1})^\top$) satisfies the equation

$$\begin{cases} \dot{e}_1 = e_2 \\ \quad\vdots \\ \dot{e}_{n-2} = e_{n-1} \\ \dot{e}_{n-1} = -\beta_1 e_1 - \beta_2 e_2 - \dots - \beta_{n-1}e_{n-1}. \end{cases}$$

It is obvious that the characteristic polynomial of this linear system coincides with $\beta_m(s)$, and, since the latter polynomial is a Hurwitz polynomial, $e' \to 0$ exponentially as $t \to \infty$. The estimate for x_n is given by $\tilde{x}_n = y - \sum_{i=1}^{n-1} \beta_i \tilde{x}_i$. Thus we have the following statement.

Theorem 5.11. *Suppose that system* (5.25) *is in the general position, has the first relative order, and is of minimal phase. Then observer* (5.31) *reconstructs the unknown part of the phase vector exponentially precisely.*

Remark 5.12. The rate of convergence of the observer is defined by the polynomial $\beta_m(s)$ and cannot be changed.

Remark 5.13. The dimension of observer (5.31) is equal to $(n-1)$ and coincides with the dimension of the Luenberger observer for fully determined systems.

Remark 5.14. In the case where no constraints are imposed on the disturbance $f(t)$, the requirement of stability of zero dynamics of system (5.25) is necessary for solving an observation problem.

Indeed, let the polynomial $\beta_m(s)$ be unstable. Let us consider a system in form (5.30) for the special case of disturbance

$$f(t) = \gamma_1 x_1 + \gamma_2 x_2 + \cdots + \gamma_{n-1} x_{n-1} + \gamma_n y.$$

Then $\dot{y} \equiv 0$, i.e., $y(t) \equiv \text{const}$. In this case, any two initial states $(x_1'(0), \dots, x_{n-1}'(0), y(0))^\top$ and $(x_2''(0), \dots, x_{n-1}''(0), y(0))$ generate the same output

$$y(t) = y(0) = \text{const}, \quad t \geq 0$$

and are indistinguishable, and the difference $e(t) = x'(t) - x''(t)$ between the solutions of the system corresponding to these initial states satisfies the equations

$$\begin{cases} \dot{e}_1 = e_2 \\ \quad \vdots \\ \dot{e}_{n-2} = e_{n-1} \\ \dot{e}_{n-1} = -\beta_1 e_1 - \cdots - \beta_{n-1} e_{n-1} \\ \dot{e}_n = 0, \end{cases}$$

and, since $\beta_m(s)$ is unstable,

$$e(t) \to \infty \quad \text{as} \quad t \to \infty.$$

Thus, two exponentially different solutions are indistinguishable.

The transformations of the system carried out in the proof of Theorem 5.11 can be conveniently written in block form, We denote by $x' = (x_1, \dots, x_{n-1})^\top$ the $(n-1)$-dimensional part of the phase vector. When we pass from the coordinates $\binom{x'}{x_n}$ to coordinates $\binom{x'}{y}$ the system assumes form (5.30) which can be written in block form

$$\begin{cases} \dot{x}' = A_{11}x' + A_{12}y \\ \dot{y} = A_{21}x' + A_{22}y + CDf, \end{cases} \tag{5.30$'$}$$

where

$$A_{11} = \begin{pmatrix} 0 & 1 & 0 & \dots & 0 \\ 0 & 0 & 1 & \dots & 0 \\ \dots & \dots & \dots & \dots & \dots \\ 0 & 0 & 0 & \dots & 1 \\ -\beta_1 & -\beta_2 & -\beta_3 & \dots & -\beta_{n-1} \end{pmatrix} \in \mathbb{R}^{(n-1)\times(n-1)},$$

$$A_{12} = \begin{pmatrix} 0 \\ \vdots \\ 0 \\ 1 \end{pmatrix} \in \mathbb{R}^{(n-1)\times 1}, \quad A_{21} = (\gamma_1, \dots, \gamma_{n-1}) \in \mathbb{R}^{1\times(n-1)}, \tag{5.32}$$

$$A_{22} = \gamma_n \in \mathbb{R}^{1\times 1}, \quad CD = 1.$$

The matrix A_{11} defines the zero dynamics of the system and in representation (5.30′) has the form of a companion matrix of the polynomial $\beta_{n-1}(s)$.

In the case of square systems with vector input f and output y (i.e., $l = m > 1$), if the system is in the general position and CD is a full-rank matrix (i.e., $\det CD \neq 0$), then, by a standard nonsingular transformation the system can also be reduced to form (5.30′) with the only difference that the matrices A_{ij} have other dimensions, namely,

$$A_{11} \in \mathbb{R}^{(n-l)\times(n-l)}, \quad A_{12} \in \mathbb{R}^{(n-l)\times l}, \quad A_{21} \in \mathbb{R}^{l\times(n-l)}, \quad A_{22} \in \mathbb{R}^{l\times l}. \tag{5.33}$$

In this case, the matrix A_{11} defines, as before, the zero dynamics of the system and, in the case of minimal phase, i.e., if A_{11} is a Hurwitz matrix, the problem is solved by the observer

$$\dot{\tilde{x}}' = A_{11}\tilde{x}' + A_{12}y \tag{5.34}$$

similar to observer (5.31) (and coinciding with it for $l = m = 1$). The dimension of this observer is equal to $(n - l)$, i.e., coincides with the dimension of the Luenberger observer. The observer reconstructs $(n - l)$ components of the phase vector, the other components in representation (5.30′), which depend explicitly on the disturbance $f(t)$, being a measurable output $y(t) \in \mathbb{R}^l$. Thus, Theorem 5.11 can be generalized to vector square systems.

Theorem 5.11′. *Suppose that system* (5.25) *is in the general position, is square (i.e., $l = m$), and is of minimal phase,* $\det CD \neq 0$. *Then observer* (5.34) *reconstructs the unknown part of the phase vector exponentially precisely.*

Scalar systems with an arbitrary relative order. In the synthesis of the observers described above an essential part is played by the nondegeneracy of the matrix CD. In the case where this matrix is degenerate, the solution of the problem becomes considerably more difficult. Let us consider in detail this case for scalar systems, i.e., where $l = m = 1$. In this case $CD \in \mathbb{R}$. The degeneracy of CD means that $CD = 0$, i.e., the relative order of the system $r > 1$. The polynomial $\beta(s)$ is of degree $(n - r)$ and the vector C has, correspondingly, the form

$$C = (\beta_1, \dots, \beta_{n-r}, \beta_{n-r+1}, 0, \dots, 0).$$

In the case where the relative order is equal to r, the first nonzero coefficient in the chain (5.29) is $CA^{r-1}D = \beta_{n-r+1}$. Without loss of generality, we assume that $CA^{r-1}D = \beta_{n-r+1} = 1$ (this can always be achieved by normalizing the output $y(t)$). Then

$$y = \beta_1 x_1 + \beta_2 x_2 + \dots + \beta_{n-r} x_{n-r} + x_{n-r+1}.$$

Suppose that system (5.25) is in the general position and is reduced to the canonical form of controllability. We carry out for it a standard transformation with the isolation of zero dynamics, for which purpose we pass from the coordinates $(x_1, \dots, x_n)^\top$ to coordinates $(x_1, \dots, x_{n-r}, y_1, \dots, y_r)^\top$, where

$$\begin{aligned} y_1 &= Cx = y \\ y_2 &= CAx = \dot{y} \\ &\vdots \\ y_r &= CA^{r-1}x = y^{(r-1)}. \end{aligned}$$

In the new coordinates the system assumes the form

$$\begin{cases} \dot{x}_1 = x_2 \\ \quad\vdots \\ \dot{x}_{n-r-1} = x_{n-2} \\ \dot{x}_{n-r} = -\beta_1 x_1 - \dots - \beta_{n-r} x_{n-r} + y \\ \dot{y}_1 = y_2 \\ \quad\vdots \\ \dot{y}_{r-1} = y_r \\ \dot{y}_r = \gamma_1' x_1 + \dots + \gamma_{n-r}' x_{n-r} + \gamma_1'' y_1 + \dots + \gamma_r'' y_r + f, \end{cases} \tag{5.35}$$

where γ_i' and γ_i'' are constants defined uniquely by the parameters of the original system.

Representation (5.35) is an analog of representation (5.30) for systems with a relative order r. For this representation we can also use the block form of notation. We denote $x' = (x_1, \dots, x_{n-r})^\top$, $y' = (y_1, \dots, y_r)^\top$, and then

$$\begin{cases} \begin{cases} \dot{x}' = A_{11}x' + A_{12}y \\ \dot{y}' = A_{21}y' + B'(f + \bar{\gamma}'x') \end{cases} \\ y = y_1 = C'y', \end{cases} \tag{5.35'}$$

where

$$A_{11} = \begin{pmatrix} 0 & 1 & 0 & \dots & 0 \\ 0 & 0 & 1 & \dots & 0 \\ \dots & \dots & \dots & \dots & \dots \\ 0 & 0 & 0 & \dots & 1 \\ -\beta_1 & -\beta_2 & -\beta_3 & \dots & -\beta_{n-r} \end{pmatrix} \in \mathbb{R}^{(n-r)\times(n-r)},$$

$$A_{12} = \begin{pmatrix} 0 \\ \vdots \\ 0 \\ 1 \end{pmatrix} \in \mathbb{R}^{(n-r)\times 1}, \quad A_{21} = \begin{pmatrix} 0 & 1 & 0 & \dots & 0 \\ 0 & 0 & 1 & \dots & 0 \\ \dots & \dots & \dots & \dots & \dots \\ 0 & 0 & 0 & \dots & 1 \\ -\gamma_1'' & -\gamma_2'' & -\gamma_3'' & \dots & -\gamma_r'' \end{pmatrix} \in \mathbb{R}^{r\times r},$$

$$B' = \begin{pmatrix} 0 \\ \vdots \\ 0 \\ 1 \end{pmatrix} \in \mathbb{R}^{r\times 1}, \quad \bar{\gamma}' = (\gamma_1', \dots, \gamma_{n-r}') \in \mathbb{R}^{1\times(n-r)},$$

$$C' = (1, 0, \dots, 0) \in \mathbb{R}^{1\times r}.$$

The matrix A_{11} defines the zero dynamics of the system and

$$\det(sI - A_{11}) = \beta_{n-r}(s) = s^{n-r} + \beta_{n-r}s^{n-r-1} + \dots + \beta_1.$$

If $\beta_{n-r}(s)$ is a Hurwitz polynomial (i.e., the system is of minimal phase), then the first $(n-r)$ unknown coordinates x' are reconstructed by the observer

$$\dot{\tilde{x}}' = A_{11}\tilde{x}' + A_{12}y \tag{5.36}$$

which differs from observer (5.34) only by the dimension. One more coordinate $y_1 = y$ is known since it coincides with the output of the system being measured. However, for $r > 1$ a problem arises of reconstruction of the remaining $(r-1)$ coordinates $y_2, \dots, y_r$ which, in fact, are the derivatives of the output of the corresponding orders. There are several approaches to the solution of this problem.

We shall describe two of them following [10] and [14].

The first approach proposed in [10] is based on the use of linear feedback with the hierarchy of amplification factors. Let us consider it in greater detail.

We shall begin with the case of a system with the maximal relative order $r = n$. In this case system (5.35′) has the form

$$\begin{cases} \dot{y}' = A_{21}y' + B'f \\ y = y_1, \end{cases} \tag{5.37}$$

the dimension of zero dynamics of the system is zero (i.e., $\beta(s) = 1$). For solving the problem we shall use a standard full-dimensional observer

$$\dot{\tilde{y}}' = A_{21}\tilde{y}' - L(C'\tilde{y}' - y), \tag{5.38}$$

the observation error $e = \tilde{y}' - y'$ satisfies the equation

$$\dot{e} = (A_{21} - LC')e - B'f = A_L e - B'f. \tag{5.39}$$

Note that the pair $\{C', A_{21}\}$ is observable, and therefore, by choosing the vector $L \in \mathbb{R}^{n\times 1}$, we can define the spectrum of the matrix A_L arbitrarily.

Let us prove an auxiliary statement.

Lemma 5.15. *Suppose that the matrix $A_L \in \mathbb{R}^{n\times n}$ is such that its spectrum can be defined arbitrarily. We denote the coefficients of decomposition of the matrix exponential by $\alpha_i(t)$, $i = 0,\dots,n-1$:*

$$e^{A_L t} = \sum_{i=0}^{n-1} \alpha_i(t) A_L^i.$$

Then, for any $\mu > 0$, the spectrum $\operatorname{Spec}\{A_L\}$ *can be chosen such that the estimate*

$$|\alpha_i(t)| \le \frac{N_i e^{-\mu t}}{\mu^i}, \quad i = 0,\dots,n-1, \tag{5.40}$$

where $N_i = \text{const} > 0$ and does not depend on μ, is valid for all $\alpha_i(t)$.

Proof. Let $\operatorname{Spec}\{A_L\} = \{\lambda_1,\dots,\lambda_n\}$. We choose λ_i such that $\lambda_i \ne \lambda_j$ for $i \ne j$. Then $\alpha_i(t)$ will satisfy the system of equations [31]

$$\sum_{i=0}^{n-1} \lambda_j^i \alpha_i(t) = e^{\lambda_j t}, \quad j = 1,\dots,n. \tag{5.41}$$

We introduce notation

$$M = \begin{pmatrix} 1 & \lambda_1 & \dots & \lambda_1^{n-1} \\ \dots & \dots & \dots & \dots \\ 1 & \lambda_n & \dots & \lambda_n^{n-1} \end{pmatrix}; \quad \alpha(t) = \begin{pmatrix} \alpha_0(t) \\ \vdots \\ \alpha_{n-1}(t) \end{pmatrix}; \quad R = \begin{pmatrix} e^{\lambda_1 t} \\ \vdots \\ e^{\lambda_n t} \end{pmatrix}.$$

Then we can write system (5.41) in the form

$$M\alpha(t) = R. \tag{5.42}$$

We choose the spectrum of A_L such that $\lambda_i = \mu\bar{\lambda}_i$, where $\bar{\lambda}_i < 0$, $|\bar{\lambda}_1| = 1$; $|\bar{\lambda}_{i+1}| > |\bar{\lambda}_i|$, $i = 1, \dots, n-1$, i.e., the spectrum of A_L is "proportional" with an amplification factor $\mu > 0$ to a certain fixed, real, distinct, and stable spectrum $\{\bar{\lambda}_1, \bar{\lambda}_2, \dots, \bar{\lambda}_n\}$. Then we can represent the matrix M in the form

$$M = \begin{pmatrix} 1 & \bar{\lambda}_1 & \dots & \bar{\lambda}_1^{n-1} \\ \dots & \dots & \dots & \dots \\ 1 & \bar{\lambda}_n & \dots & \bar{\lambda}_n^{n-1} \end{pmatrix} \begin{pmatrix} 1 & 0 & \dots & 0 \\ 0 & \mu & \dots & 0 \\ \dots & \dots & \dots & \dots \\ 0 & 0 & \dots & \mu^{n-1} \end{pmatrix} = \bar{M} D_\mu.$$

Solving equation (5.42), we obtain

$$\begin{aligned} \alpha(t) &= M^{-1}R = (\bar{M}D_\mu)^{-1}R = D_\mu^{-1}\bar{M}^{-1}R \\ &= \begin{pmatrix} 1 & 0 & \dots & 0 \\ 0 & \dfrac{1}{\mu} & \dots & 0 \\ \dots & \dots & \dots & \dots \\ 0 & 0 & \dots & \dfrac{1}{\mu^{n-1}} \end{pmatrix} \bar{M}^{-1}R. \end{aligned} \tag{5.43}$$

Since $e^{\lambda_i t} = e^{\bar{\lambda}_i \mu t} = \left(e^{-\mu t}\right)^{|\bar{\lambda}_i|}$, and $|\bar{\lambda}_i| \geq 1$ for $i = 1, \dots, n$, the estimate

$$|R(t)| \leq r_0 e^{-\mu t}, \quad t > 0,$$

is obvious for the vector R. Taking into account this inequality and also the fact that the matrix $\bar{M}^{-1}$ does not depend on μ, we find from (5.43) that

$$|\alpha_i(t)| = \frac{1}{\mu^i}|\bar{M}_i^{-1}R| \leq \frac{e^{-\mu t}}{\mu^i} r_0 |\bar{M}_i^{-1}| = \frac{N_i e^{-\mu t}}{\mu^i},$$

where $\bar{M}_i^{-1}$ is the ith row of the matrix $\bar{M}^{-1}$. The lemma is proved. □

For solving an observation problem, we shall use, as was indicated above, a standard full-dimensional observer (5.38) and choose the spectrum of the matrix A_L in accordance with Lemma 5.15, i.e.,

$$\begin{gathered} \operatorname{Spec}\{A_L\} = \{\mu\bar{\lambda}_1, \dots, \mu\bar{\lambda}_n\}, \quad \bar{\lambda}_i < 0, \quad i = 1, \dots, n, \quad \mu > 0, \\ |\bar{\lambda}_1| = 1, \quad |\bar{\lambda}_{i+1}| > |\bar{\lambda}_i|, \quad i = 1, \dots, n-1. \end{gathered} \tag{5.44}$$

Then the following statement holds true.

Theorem 5.16. *Suppose that system* (5.25) *is in the general position, its relative order* $r = n$, *the system is reduced to the canonical form* (5.37). *Suppose, in addition, that the unknown input* $f(t)$ *is uniformly bounded by the known constant* F_0, *i.e.,* $|f(t)| \leq F_0$ *for* $t \geq 0$.

We choose the feedback vector L *in observer* (5.38) *such that the spectrum of the matrix* A_L *should satisfy conditions* (5.44). *Then the observation error* $e = \tilde{y}' - y'$ *satisfies the estimate*

$$|e(t)| \leq K_1 e^{-\mu t} + \frac{K_2}{\mu}, \tag{5.45}$$

where the constant K_2 *does not depend on the amplification factor* μ.

Proof. As was shown above, when we use observer (5.38), the observation error satisfies equation (5.39), i.e.,

$$\dot{e} = A_L e - B' f,$$

and we can find $e(t)$ using the Cauchy formula for solving the linear equation

$$e(t) = e(0)e^{A_L t} - \int_0^t e^{A_L(t-\tau)} B' f(\tau)\, d\tau = e(0)e^{A_L t} - \int_0^t e^{A_L \tau} B' f(t-\tau)\, d\tau.$$

Let us estimate the norm of the vector $e(t)$

$$|e(t)| \leq \left|e(0)e^{A_L t}\right| + \left|\int_0^t e^{A_L \tau} B' f(t-\tau)\right| d\tau.$$

We choose L such that the spectrum of A_L should satisfy conditions (5.44). In this case A_L is a Hurwitz matrix and the estimate

$$|e^{A_L t}| \leq K_0 e^{-\mu t},$$

where $K_0 = \text{const} > 0$ (generally speaking, depends on μ), holds for $e^{A_L t}$. Moreover, the expansion

$$e^{A_L \tau} = \sum_{i=0}^{n-1} \alpha_i(\tau) A_L^i,$$

where the functions $\alpha_i(\tau)$ satisfy estimates (5.40), holds for the matrix exponential (by virtue of the choice of the spectrum of A_L). Then we have an estimate

$$\begin{aligned} |e(t)| &\leq |e(0)| K_0 e^{-\mu t} + \left|\int_0^t \sum_{i=0}^{n-1} \alpha_i(\tau) A_L^i B' f(t-\tau)\right| d\tau \\ &\leq |e(0)| K_0 e^{-\mu t} + \sum_{i=0}^{n-1} \left[\int_0^t |\alpha_i(\tau)| \cdot |A_L^i B'| \cdot |f(t-\tau)|\right] d\tau \end{aligned}$$

for $|e(t)|$.

Let us consider in greater detail the structure of the matrices $A_L^i B'$. In accordance with (5.35′), the explicit representations of the matrices are given:

$$B' = \begin{pmatrix} 0 \\ \vdots \\ 0 \\ 1 \end{pmatrix}, \quad A_L = \begin{pmatrix} 0 & 1 & 0 & \dots & 0 \\ 0 & 0 & 1 & \dots & 0 \\ \dots & \dots & \dots & \dots & \dots \\ 0 & 0 & 0 & \dots & 1 \\ -\gamma_1'' & -\gamma_2'' & -\gamma_3'' & \dots & -\gamma_n'' \end{pmatrix} - \begin{pmatrix} l_1(\mu) & 0 & \dots & 0 \\ l_2(\mu) & 0 & \dots & 0 \\ \dots & \dots & \dots & \dots \\ l_n(\mu) & 0 & \dots & 0 \end{pmatrix}.$$

Here $l_i(\mu)$ are components of the feedback vector L; note that only they depend on the choice of μ. It follows from the form of the matrix A_L that $(A_L)^0 = I_n$ does not depend on μ and in matrix $(A_L)^1$ only the first column depends on μ. Let us consider in greater detail the square $(A_L)^2$. We have

$$(A_L)^2 = \begin{pmatrix} -l_1(\mu) & 1 & 0 & \dots & 0 \\ -l_2(\mu) & 0 & 1 & \dots & 0 \\ \dots & \dots & \dots & \dots & \dots \\ -l_{n-1}(\mu) & 0 & 0 & \dots & 1 \\ (-l_n(\mu) - \gamma_1'') & -\gamma_2'' & -\gamma_3'' & \dots & -\gamma_n'' \end{pmatrix}$$

$$\times \begin{pmatrix} -l_1(\mu) & 1 & 0 & \dots & 0 \\ -l_2(\mu) & 0 & 1 & \dots & 0 \\ \dots & \dots & \dots & \dots & \dots \\ -l_{n-1}(\mu) & 0 & 0 & \dots & 1 \\ (-l_n(\mu) - \gamma_1'') & -\gamma_2'' & -\gamma_3'' & \dots & -\gamma_n'' \end{pmatrix}$$

$$= \begin{pmatrix} * & -l_1(\mu) & 1 & 0 & \dots & 0 \\ * & -l_2(\mu) & 0 & 1 & \dots & 0 \\ \dots & \dots & \dots & \dots & \dots & \dots \\ * & -l_{n-2}(\mu) & 0 & 0 & \dots & 1 \\ * & (-l_{n-1}(\mu) - \gamma_2'') & -\gamma_3'' & -\gamma_4'' & \dots & -\gamma_{n-1}'' \\ * & (-l_n(\mu) - \gamma_1'' + \gamma_2''\gamma_n'') & (-\gamma_2'' + \gamma_3''\gamma_n'') & (-\gamma_3'' + \gamma_4''\gamma_n'') & \dots & (-\gamma_{n-1}'' + \gamma_n''\gamma_n'') \end{pmatrix}$$

$$= \begin{pmatrix} * & * & 1 & 0 & \dots & 0 \\ * & * & 0 & 1 & \dots & 0 \\ \dots & \dots & \dots & \dots & \dots & \dots \\ * & * & 0 & 0 & \dots & 1 \\ & * & -\gamma_3'' & -\gamma_4'' & \dots & -\gamma_{n-1}'' \\ * & * & (-\gamma_2'' + \gamma_3''\gamma_n'') & (-\gamma_3'' + \gamma_4''\gamma_n'') & \dots & (-\gamma_{n-1}'' + \gamma_n''\gamma_n''), \end{pmatrix},$$

where the sign $*$ denotes elements dependent on μ. It follows that in the matrices $(A_L)^2$ only the first two columns depend on μ. Continuing explicit computations, we find that in the matrices A_L^i only the first i columns depend on μ ($i = 1, \dots, n-1$). The multiplication of the matrix by the column B' gives the last column of the matrix. Since for all $i = 1, \dots, n-1$ the last columns of the matrix A_L^i do not depend on μ, it follows that

$$|A_L^i B'| = Q_i,$$

where Q_i are constants which are independent of μ and are defined only by the parameters of the system. Taking this fact into account, as well as the uniform boundedness of $f(t)$ and estimates (5.40) for the functions $\alpha_i(t)$, we have the following estimate for the measurement error $e(t)$:

$$|e(t)| \le |e(0)|K_0 e^{-\mu t} + \sum_{i=0}^{n-1} Q_i F_0 \frac{N_i}{\mu^i} \int_0^t e^{-\mu\tau} d\tau$$

$$= |e(0)|K_0 e^{-\mu t} + \sum_{i=0}^{n-1} \frac{F_0 N_i Q_i}{\mu^i} \frac{(1-e^{\mu t})}{\mu} \le |e(0)|K_0 e^{-\mu t} + \frac{1}{\mu} \sum_{i=0}^{n-1} \frac{F_0 N_i Q_i}{\mu^i}.$$

Without loss of generality, we assume that $\mu \ge \mu^* > 0$. Then

$$\sum_{i=0}^{n-1} \frac{F_0 N_i Q_i}{\mu^i} \le \sum_{i=0}^{n-1} \frac{F_0 N_i Q_i}{(\mu^*)^i} = K_2,$$

where the constant K_2 does not depend on the choice of the amplification factor μ. We denote $|e(0)|K_0 = K_1$ and find the final estimate

$$|e(t)| \le K_1 e^{-\mu t} + \frac{K_2}{\mu}.$$

The theorem is proved. □

Corollary 5.17. *It follows from estimate* (5.45) *that when defining the spectrum of the matrix A_L from condition* (5.44)*, by the choice of a sufficiently large factor $\mu > 0$ we can make the estimation error in the asymptotics smaller than any preassigned value.*

Corollary 5.18. *The constant K_1 in estimate* (5.45) *depends on the unknown initial deviation $e(0)$ and on μ. Moreover, $K_1 \to \infty$ as $\mu \to \infty$.*

Therefore observers (5.39) for large values of the factor μ are characterized by an "initial burst" when the error "quickly" increases at the initial moment and then "quickly" decreases, and then keeps in the given range.

Corollary 5.19. *Conditions* (5.44) *establish the hierarchy of the coefficients of the feedback matrix L according to the degrees of the amplification factor μ. The coefficients $l_i(\mu)$ have the simplest form in the case where all $\gamma_i'' = 0$, i.e., A_{21} from* (5.37) *has the form*

$$A_{21} = \begin{pmatrix} 0 & 1 & 0 & \dots & 0 \\ 0 & 0 & 1 & \dots & 0 \\ \dots & \dots & \dots & \dots & \dots \\ 0 & 0 & 0 & \dots & 1 \\ 0 & 0 & 0 & \dots & 0 \end{pmatrix}. \quad (5.46)$$

Then

$$A_L = A_{21} - LC' = \begin{pmatrix} -l_1(\mu) & 1 & 0 & \dots & 0 \\ -l_2(\mu) & 0 & 1 & \dots & 0 \\ \dots & \dots & \dots & \dots & \dots \\ -l_{n-1}(\mu) & 0 & 0 & \dots & 1 \\ -l_n(\mu) & 0 & 0 & \dots & 0 \end{pmatrix},$$

$$\det(sI - A_L) = s^n + l_1(\mu)s^{n-1} + \dots + l_n(\mu).$$

We choose a set $\bar{\lambda}_i$ from condition (5.44) This set is associated with the Hurwitz polynomial

$$\bar{\varphi}(s) = \prod_{i=1}^{n}(s - \bar{\lambda}_i) = s^n + \bar{l}_1 s^{n-1} + \dots + \bar{l}_n. \tag{5.47}$$

Then the set $\mu\bar{\lambda}_1, \dots, \mu\bar{\lambda}_n$ is associated with the polynomial

$$\varphi(s) = \prod_{i=1}^{n}(s - \mu\lambda_i) = s^n + \mu\bar{l}_1 s^{n-1} + \mu^2\bar{l}_2 s^{n-2} + \dots + \mu^n\bar{l}_n.$$

Thus, for $\operatorname{spec}\{A_L\} = \{\mu\bar{\lambda}_1, \dots, \mu\bar{\lambda}_n\}$ for the matrix A_{21} from (5.46), the feedback vector L in observer (5.38) should be chosen in the form

$$L = (\mu\bar{l}_1, \mu^2\bar{l}_2, \dots, \mu^n\bar{l}_n)^\top, \tag{5.48}$$

whence we see the hierarchy of the coefficients $l_i(\mu)$ according to the degrees of the parameter μ mentioned above.

If the matrix A_{21} has a more complicated structure, i.e., $\gamma_i'' \neq 0$, then the feedback vector L will also have a form more complicated than (5.48) We can easily verify that in this case

$$l_i(\mu) = \mu^i\bar{l}_i + \tilde{l}_i(\mu) = \mu^i\bar{l}_i + o(\mu^i),$$

where $\tilde{l}_i(\mu)$ is a polynomial of μ of a degree lower than i, i.e., $l_i(\mu)$ is a polynomial of μ of a degree equal to i with the leading coefficient $\bar{l}_i$. The other terms, the polynomials $\tilde{l}_i(\mu)$, can be found in explicit form from the coefficients γ_j''.

In practical applications, for simplicity, in observer (5.38) and for $\gamma_i'' \neq 0$ we can use the vector L from (5.48). In this case $\det(sI - A_L)$ does not coincide with the required polynomial $\gamma(s)$. However, as $\mu \to \infty$, the spectrum of A_L (with L from (5.48)) tends to the given spectrum from (5.44), i.e., for $\mu \geq \mu^* = \text{const}$ the spectrum of A_L becomes real and distinctive and, consequently, the asymptotic estimate (5.45) is valid for the observation error.

5.5.2 Observers for systems with arbitrary relative order $r > 1$

The proposed approach to the construction of observers can be generalized to systems with an arbitrary relative order.

Suppose that the relative order of the system is equal to r, where $1 < r < n$. Then, by a nonsingular transformation, the system can be reduced to form (5.35′)

$$\begin{cases} \begin{cases} \dot{x}' = A_{11}x' + A_{12}y' \\ \dot{y}' = A_{21}y' + B'(f + \bar{\gamma}'x') \end{cases} \\ y = y_1 = C'y', \end{cases}$$

where a part of the phase vector $x' \in \mathbb{R}^{n-r}$ corresponds to the zero dynamics of the system. If A_{11} is a Hurwitz matrix (i.e., the characteristic polynomial of zero dynamics $\beta_{n-r}(s) = \det(sI - A_{11})$ is a Hurwitz polynomial, the system is of minimal phase), then observer (5.36) reconstructs a part of the phase vector x' exponentially.

To reconstruct y' we use the method which was proposed when we contracted an observer for systems with maximal relative order, namely, we use an observer of the form

$$\dot{\tilde{y}}' = A_{21}\tilde{y}' - L(C'\tilde{y}' - y) + B'\bar{\gamma}'\tilde{x}', \tag{5.49}$$

where $\tilde{x}'$ is the estimate of x' from observer (5.36). The estimation error $e(t) = (\tilde{y}' - y') \in \mathbb{R}^r$ satisfies the equation

$$\dot{e} = A_L e - B'(f - \bar{\gamma}'e')$$

which differs from equation (5.39) by its dimension and the presence of an exponentially decreasing term $e'(t) = \tilde{x}' - x'$, i.e., the error of observer (5.36). By analogy with Theorem 5.16, we choose a vector $L \in \mathbb{R}^{r\times 1}$ such that

$$\operatorname{Spec}\{A_L\} = \{\mu\bar{\lambda}_1, \ldots, \mu\bar{\lambda}_r\}, \quad \bar{\lambda}_i < 0, \ i = 1, \ldots, r, \quad \mu > 0,$$
$$|\bar{\lambda}_1| = 1, \ |\bar{\lambda}_{i+1}| > |\bar{\lambda}_i|, \ i = 1, \ldots, r-1.$$

Then, carrying out estimates as we did in the proof of Theorem 5.16, we obtain an estimate for the observation error $e(t)$

$$|e(t)| \le K_1 e^{-\mu t} + \frac{K_2}{\mu} + K_3 e^{-\delta t},$$

where, as before, $K_2 = \text{const}$ does not depend on μ and the constant $\delta > 0$ characterizes the degree of stability of the polynomial $\beta_{n-r}(s)$ (i.e., the degree of stability of the zero dynamics of the system)

$$\operatorname{Spec}\{A_{11}\} = \{\delta_1, \ldots, \delta_r\}, \quad \operatorname{Re}(\delta_i) < -\delta, \ i = 1, \ldots, r.$$

Consequently, choosing the parameter $\mu > 0$ sufficiently large, we can make in asymptotics an estimation error $e(t)$ smaller than any preassigned value.

Observers (5.36) and (5.49) together form an observer of full order n for a minimal-phase system with an arbitrary relative order. In this case, a part of the phase vector is reconstructed exponentially precisely and another part with the preassigned accuracy. All corollaries of Theorem 5.16 which were made for systems with the maximal relative order are valid for observer (5.49).

Conclusion

In Chap. 5 we considered the problem of synthesis of asymptotic observers for linear stationary systems under the conditions of uncertainty. When solving this problem we can distinguish two cases, namely, hyperoutput systems (when the dimension of the output exceeds that of the unknown input) and square systems (when these dimensions coincide).

For hyperoutput systems two approaches to solving the problem are proposed, namely, the method of quasisplitting and the method of pseudoinputs. These two methods allow us to obtain a solution of the problem under the same requirements set for the system. The main results are formulated in Theorem 5.1 and Theorem 5.3, respectively. The proposed methods make it possible to solve a number of auxiliary problems, in particular, the method of quasisplitting allow us to solve the problem of synthesis of functional observers (Sec. 5.2) and the method of pseudoinputs allow us to obtain a number of representations which are convenient for solving problems of stabilization and observation under the conditions of uncertainty (Theorem 5.3′ and Theorem 5.3″).

In Sec. 5.5 we consider the problem of synthesis of observers for square systems. We propose an approach to the solution of this problem based on the hierarchy of feedback amplification factors which makes it possible to obtain an estimate of the unknown phase vector with any preassigned accuracy, but this approach is not asymptotic (Theorem 5.16).

Chapter 6

Observers for bilinear systems

In this chapter we consider a problem of constructing observers for one of the classes of nonlinear dynamical systems, namely, for bilinear systems of the form

$$\begin{cases} \dot{x} = Ax + uBx + vD \\ y = Cx, \end{cases} \tag{6.1}$$

where, as before, $x \in \mathbb{R}^n$ is an unknown phase vector of the system, $u \in \mathbb{R}^m$ and $v \in \mathbb{R}^k$ are known inputs of the system, and $y \in \mathbb{R}^l$ is the measurable output; $A,\ B,\ C,\ D$ are known constant matrices of the corresponding dimensions.

The necessity of considering observation problems for this class of systems is explained by the fact that upon the degeneration of the bilinearity matrices, i.e., the matrix B in (6.1), and the linear output, the synthesis of state observers and, correspondingly, functional observers can be carried out by the methods proposed in the preceding chapters. In the absence of this degeneration the observation problem for systems of form (6.1) is rather difficult. This is demonstrated by an example of planar (i.e., for $n = 2$) bilinear systems.

Since the effect produced by the known input $v(t)$ can always be compensated in the observer, in what follows we shall consider bilinear systems of the form

$$\begin{cases} \dot{x} = Ax + uBx \\ y = Cx. \end{cases} \tag{6.2}$$

The singularity of this system consists in the fact that for any matrices $A,\ B$, and C this system loses controllability at the point $x = 0$.

6.1 Asymptotic observers of bilinear systems in the plane

Problem statement. We consider a standard observation problem for the bilinear dynamical system (6.2), where $x \in \mathbb{R}^2$, A and $B \in \mathbb{R}^{2\times 2}$ are known matrices with constant coefficients; $u \in \mathbb{R}$ is a scalar and, when necessary, known input function of the system. We have to formulate an exponential estimate $\hat{x}(t)$ of the phase vector $x(t)$ of system (6.2) using the continuous measurements of the scalar output $y = Cx$, where $C \in \mathbb{R}^{1\times 2}$. This problem is a classical observation problem for controllable dynamical systems. The complexity of its solution for a bilinear system in due to the presence, on the right-hand side, generally speaking, of an alternating function $u(t)$. We shall explain this fact. When we use the standard Luenberger observer

$$\dot{\hat{x}} = A\hat{x} + uB\hat{x} - l(C\hat{x} - y) - ug(C\hat{x} - y), \tag{6.3}$$

where l and g are feedback parameters, the estimation problem reduces to the stability of the system in deviations $e = \hat{x} - x$

$$\dot{e} = A_l e + uB_g e, \tag{6.4}$$

where $A_l = A - lC$, $B_g = B - gC$, and the fact that A_l and B_g are Hurwitz matrices in the variable $u(t)$, does not imply, in general, the stability of the linear system (6.4) with matrix $A_l + u(t)B_g$. It is shown in [8] that it suffices for $A_l = A - lC$ to be a Hurwitz matrix for the stability of system (6.4) for sufficiently small controls $u(t)$, but for "large", even for uniformly bounded with respect to t, functions $u(t)$, the Luenberger observer (6.3) does not already solve the problem. Therefore we should find the means of eliminating the effect produced by the function $u(t)$ on the robustness of the estimation processes.

Let us consider the algorithms of synthesizing the state observers of system (6.2) which solve the problem exponentially precisely (as close to the preassigned degree of stability as possible) for any uniformly, with respect to t, bounded input functions, i.e., under the condition

$$|u(t)| \le u_0, \quad t \ge 0. \tag{6.5}$$

For our purpose we shall prove some auxiliary statements. For the two-dimensional bilinear system (6.2), without loss of generality, we shall consider, as can be assumed, that rank $B = 1$, i.e., $B = bh$, where b and h are known vector column and vector row, respectively.

Indeed, we take a row h such that $\det\begin{pmatrix} h \\ C \end{pmatrix} \neq 0$. Let b^1 and b^2 be columns of the matrix $\hat{B} = \begin{pmatrix} h \\ C \end{pmatrix} B \begin{pmatrix} h \\ C \end{pmatrix}^{-1}$. Denoting $b' = \begin{pmatrix} h \\ C \end{pmatrix}^{-1} \cdot b^1$, $b'' = \begin{pmatrix} h \\ C \end{pmatrix}^{-1} \cdot b^2$, we rewrite system (6.2) in the form

$$\dot{x} = Ax + ub'hx + ub''y.$$

In this case, using the auxiliary system

$$\dot{\tilde{x}} = A\tilde{x} + ub'h\tilde{x} + ub''y \tag{6.6}$$

we reduce the problem of estimating the phase vector $x = \tilde{x} + e$ to the observation problem for the system in deviations

$$\dot{e} = Ae + u\tilde{B}e, \quad \tilde{B} = b'h, \tag{6.7}$$

where rank $\tilde{B} = 1$.

To make the further references more convenient, we give the following statement.

Lemma 6.1. *The relation*

$$\det\begin{pmatrix} C \\ CA_u \end{pmatrix} = \det\begin{pmatrix} C \\ CA \end{pmatrix} + u \det\begin{pmatrix} C \\ CB \end{pmatrix} \tag{6.8}$$

holds for the arbitrary (2×2) *matrices* $A_u = A + uB$ *and the row* $C \in R^{1\times 2}$.
If, in addition, the factorization $B = bh$ *is valid, then*

$$\det\begin{pmatrix} C \\ CA_u \end{pmatrix} = \det\begin{pmatrix} C \\ CA \end{pmatrix} + u(Cb) \det\begin{pmatrix} C \\ h \end{pmatrix}. \tag{6.9}$$

The *proof* can be carried out by a direct verification.
The following lemma is useful for synthesizing observers.

Lemma 6.2. *Suppose that the triple* $\{C, A, b\}$ *is in the general position*[1] *and the condition* $\det A_{\lambda^*} = \det(\lambda^* I - A) \neq 0$ *is fulfilled for*

$$\lambda = \lambda^* = \frac{CA^{-1}b}{Cb} \det A. \tag{6.10}$$

Then the row

$$d_* = CA_{\lambda^*}^{-1} \tag{6.11}$$

is orthogonal to the vector b, *i.e.,* $d_* b = 0$, *and, in addition,* $\det\begin{pmatrix} C \\ d_* \end{pmatrix} \neq 0$.

Proof. By virtue of the Hamilton–Cayley theorem we have

$$A^2 - A \operatorname{tr} A + I \det A = 0.$$

Then

$$A - I \operatorname{tr} A + A^{-1} \det A = 0$$

for any nondegenerate matrix A including A_λ for all $\lambda \notin \operatorname{spec}\{A\}$. Let us premultiply this identity by C and postmultiply it by b for A_λ,

$$CA_\lambda b - Cb \operatorname{tr} A_\lambda - CA_\lambda^{-1} \det A_\lambda b = 0.$$

Note that by virtue of the choice of λ^* from (6.10) we have

$$\begin{aligned} CA_\lambda^* b - Cb \operatorname{tr} A_{\lambda^*} &= C(\lambda^* I - A)b - Cb \operatorname{tr}(\lambda^* I - A) \\ &= Cb\lambda^* - CAb - Cb2\lambda^* + Cb \operatorname{tr} A = -CAb + Cb \operatorname{tr} A - Cb\lambda^* \\ &= C(-A + I \operatorname{tr} A - A^{-1} \det A)b = 0 \end{aligned}$$

and, hence, $d_* b = 0$, where d_* is defined in (6.11).

[1] Note that we speak about two-dimensional vectors and (2×2) matrices, and, in addition, $\det A \neq 0$, $Cb \neq 0$, $\det\begin{pmatrix} C \\ CA \end{pmatrix} \neq 0$, and $\det(b, Ab) \neq 0$.

For proving the second statement we should note that $d_* = CA_{\lambda^*}^{-1} = \frac{1}{\det A^*}(CA - C\,\mathrm{tr}\,A)$, and therefore

$$\det\begin{pmatrix} C \\ d_* \end{pmatrix} = \frac{1}{\det A_{\lambda^*}} \det\begin{pmatrix} C \\ CA \end{pmatrix}$$

and, by virtue of the conditions of the lemma, $\det\begin{pmatrix} C \\ d_* \end{pmatrix} \neq 0$. Lemma 6.2 is proved.□

Remark 6.3. If the triple $\{C, A, b\}$ is in the general position, then we can introduce for it a transfer function

$$W(s) = C(sE - A)^{-1}b = \frac{\beta(s)}{\alpha(s)}$$

with $\deg(\beta(s)) = 1$ in the conditions of the lemma. In this case λ^* is the zero of the transfer function, i.e., $\beta(\lambda^*) = W(\lambda^*) = 0$.

Indeed, under the conditions of Lemma 6.2

$$\beta(s) = \beta_2 s + \beta_1,\ \alpha(s) = s^2 + \alpha_2 s + \alpha_1,$$

and $\alpha_1 = \det A$, $\alpha_2 = -\,\mathrm{tr}\,A$ but $\beta_2 = Cb$. In addition, $W(0) = \beta_1/\alpha_1 = -CA^{-1}b$. It follows immediately that $\beta_1 = -CA^{-1}b \cdot \det A$ and the root of the numerator $\beta(s)$ is defined by the relation

$$s = -\frac{\beta_1}{\beta_2} = \frac{CA^{-1}b}{Cb} \det A = \lambda^*.$$

Conditions of observability uniform with respect to t. The system under consideration is observable uniformly with respect to $t \geq 0$ if

$$\min_{t\geq 0}\left|\det\begin{pmatrix} C \\ C(A + u(t)B) \end{pmatrix}\right| > 0. \tag{6.12}$$

Since

$$\det\begin{pmatrix} C \\ CA_u \end{pmatrix} = \det\begin{pmatrix} C \\ CA \end{pmatrix} + uCb \det\begin{pmatrix} C \\ h \end{pmatrix}$$

according to Lemma 6.1, where $u(t)$, in general, is an arbitrary uniformly bounded function (satisfying condition (6.5)), we can easily make sure that the following statement is valid.

Theorem 6.4. *The bilinear system $\dot{x} = Ax + ubhx$ in the plane with a scalar output $e = Cx$ and an arbitrary bounded function $|u(t)| \leq u_0$ is uniformly, with respect to t, observable if and only if*

$$\det\begin{pmatrix} C \\ CA \end{pmatrix} + u(t)Cb \det\begin{pmatrix} C \\ h \end{pmatrix} \neq 0,\quad t \geq 0.$$

From this statement we can obtain sufficient and necessary conditions of uniform observability.

Lemma 6.5.

1°. *The pair $\{C, A + ubh\}$, where $|u(t)| \le u_0$, is observable uniformly with respect to t if one of the following conditions is fulfilled:*

(a) $Cb \neq 0$, $\det\begin{pmatrix} C \\ h \end{pmatrix} \neq 0$, $\left|\det\begin{pmatrix} C \\ CA \end{pmatrix}\right| > u_0|Cb|\left|\det\begin{pmatrix} C \\ h \end{pmatrix}\right|$,

(b) *the pair $\{C, A\}$ is observable and the vectors C and h are collinear,*

(c) *the pair $\{C, A\}$ is observable and $Cb = 0$;*

2°. *If the pair $\{C, A + ubh\}$, where $|u(t)| \le u_0$, is observable uniformly with respect to t, then the pair $\{C, A\}$ is observable.*

Thus, in the case of the general position, the uniform observability is guaranteed if the constraint

$$u_0 < \left| \frac{\det\begin{pmatrix} C \\ CA \end{pmatrix}}{Cb \det\begin{pmatrix} C \\ h \end{pmatrix}} \right| \tag{6.13}$$

is fulfilled. In what follows, we shall call this condition by a condition of strict uniform observability. If the problem is degenerate (conditions (b) and (c) of Lemma 6.5), then the constraint imposed on the control $u(t)$ is absent.

State observers for degenerate bilinear systems in the plane. We shall begin the consideration with the simplest case (b) of Lemma 6.5 where $\det\begin{pmatrix} C \\ h \end{pmatrix} = 0$. Without loss of generality, we assume that $y = z = hx$ (this can be achieved by normalization of the input). In this case the problem is solved by an observer of the form

$$\dot{\hat{x}} = A\hat{x} + uby - l(C\hat{x} - y), \tag{6.14}$$

where $\hat{x}$ is the estimate of the phase vector, l is a feedback vector defined arbitrarily. With due account of the relation $y = z$ the estimation error $\varepsilon = \hat{x} - x$ satisfies the equation

$$\dot{\varepsilon} = A_l \varepsilon, \quad A_l = A - lC. \tag{6.15}$$

If the pair $\{C, A\}$ is observable, then, by choosing the vector l, we can define the spectrum of the matrix A_l arbitrarily, and, consequently, the following theorem is valid.

Theorem 6.6. *Suppose that the following conditions are fulfilled in the bilinear system $\dot{x} = Ax + ubhx$ in the plane with a scalar output $y = Cx$:*

(1) $\det\begin{pmatrix} C \\ h \end{pmatrix} = 0$,

(2) *the pair $\{C, A\}$ is observable.*

Then observer (6.14) *solves exponentially precisely the observation problem with any preassigned exponent for any function $u(t)$.*

Remark 6.7. Generally speaking, there is no necessity in a full-dimensional observer of form (6.14), the consideration can be restricted to an observer of the first order. Somewhat later, when analyzing another degenerate case, we shall describe the technique of constructing such an observer of lowered order.

Let us consider case (c) from Lemma 6.5. In this situation $Cb = 0$ and it is convenient to pass from the original variables x to a space of output derivatives $(y, \dot{y})$. Setting $y_1 = y$, $y_2 = \dot{y}$, we find an algebraic connection between the new and the old variables in the form

$$\begin{pmatrix} y_1 \\ y_2 \end{pmatrix} = \begin{pmatrix} C \\ CA \end{pmatrix} x.$$

If the pair $\{C, A\}$ is observable, as is assumed in what follows, we can take

$$\hat{x} = \begin{pmatrix} C \\ CA \end{pmatrix}^{-1} \begin{pmatrix} y_1 \\ \hat{y}_2 \end{pmatrix}, \tag{6.16}$$

where $\hat{y}_2$ is the estimate of the variable y_2, as the estimate of $\hat{x}$.

In order to solve the problem, we find the equation of the bilinear system under consideration in coordinates (y_1, y_2). We have

$$\dot{y}_2 = \ddot{y} = CA\dot{x} = CA^2x + uCAbhx$$

(it is obvious that $CAb \neq 0$ under the assumptions that we have made, since, otherwise, $b = 0$). Since $A^2 = A \operatorname{tr} A - I \det A$, it follows that

$$\dot{y}_2 = -y_1 \det A + y_2 \operatorname{tr} A + uCAbh \begin{pmatrix} C \\ CA \end{pmatrix}^{-1} \begin{pmatrix} y_1 \\ y_2 \end{pmatrix}.$$

To make the further computations simpler, we introduce notation

$$CAbh \begin{pmatrix} C \\ CA \end{pmatrix}^{-1} = (a_1, a_2). \tag{6.17}$$

With this notation, the required system assumes the form

$$\begin{cases} \dot{y}_1 = y_2 \\ \dot{y}_2 = (ua_1 - \det A)y_1 + (\operatorname{tr} A + a_2 u)y_2, \end{cases} \tag{6.18}$$

with the observation equation

$$y = y_1. \tag{6.19}$$

According to the traditions of a linear control theory, we take for system (6.18), (6.19) an observer in the form

$$\begin{cases} \dot{\hat{y}}_1 = \hat{y}_2 - k_2(\hat{y}_1 - y) \\ \dot{\hat{y}}_2 = (ua_1 - \det A)y + (\operatorname{tr} A + a_2 u)\hat{y}_2 - k_1(\hat{y}_1 - y), \end{cases} \tag{6.20}$$

where k_1 and k_2 are the adjustable feedback parameters of the observer. However, such an approach leads to a rather difficult problem of analysis of the stability of the nonstationary system in deviations[2] $\varepsilon_1 = \hat{y}_1 - y_1$, $\varepsilon_2 = \hat{y}_2 - y_2$

$$\begin{cases} \dot{\varepsilon}_1 = \varepsilon_2 - k_2\varepsilon_1 \\ \dot{\varepsilon}_2 = (\operatorname{tr} A + a_2 u)\varepsilon_2 - k_1\varepsilon_1. \end{cases} \tag{6.21}$$

Instead, we shall try to synthesize an observer of a lowered order, one-dimensional in this case, and this will make essentially simpler the synthesis of the observer and the analysis of its properties. For convenience, we shall rewrite (6.18) as

$$\begin{cases} \dot{y}_1 = y_2 \\ \dot{y}_2 = (\alpha_1 + a_1 u)y_1 + (\alpha_2 + a_2 u)y_2. \end{cases} \tag{6.22}$$

Here we assume that $\alpha_1 = -\det A$, $\alpha_2 = \operatorname{tr} A$. We introduce a new variable

$$\sigma = y_2 + dy_1, \tag{6.23}$$

where d is a constant, different from zero, whose choice will be indicated below. Let us pass to the new variables (y_1, σ). Then we have

$$\dot{\sigma} = [(\alpha_1 + a_1 u) - (\alpha_2 + a_2 u + d)d]y_1 + (\alpha_2 + a_2 u + d)\sigma.$$

Let us denote $\beta_1 = (\alpha_1 + a_1 u) - (\alpha_2 + a_2 u + d)d$, $\beta_2 = \alpha_2 + a_2 u + d$. In this notation the equations of the system being estimated in the variables (y_1, σ) assume the form

$$\begin{cases} \dot{y}_1 = -dy_1 + \sigma \\ \dot{\sigma} = \beta_1 y_1 + \beta_2\sigma, \quad y = y_1. \end{cases} \tag{6.24}$$

Since the variable $y_1 = y$ is known, it suffices to estimate only the variable σ. This problem for $\beta_2 < 0$ is solved by the scalar observer

$$\dot{\hat{\sigma}} = \beta_1 y + \beta_2\hat{\sigma}. \tag{6.25}$$

Indeed, in this case the estimation error $\varepsilon = \hat{\sigma} - \sigma$ satisfies the scalar equation

$$\dot{\varepsilon} = \beta_2\varepsilon$$

which is exponentially stable with the defined exponent $\eta > 0$ if (by virtue of uniform boundedness of $u(t)$)

$$d \le -|a_2|u_0 - \operatorname{tr} A - \eta. \tag{6.26}$$

To estimate the original phase vector x we use formula (6.16) which, with due account of the introduced notation, assumes the form

$$\hat{x} = \begin{pmatrix} C \\ CA \end{pmatrix}^{-1} \begin{pmatrix} y \\ \hat{\sigma} - dy \end{pmatrix}. \tag{6.27}$$

Thus, the following theorem is proved.

[2]Later we shall give special attention to this problem.

Theorem 6.8. *Suppose that the following conditions are fulfilled in the bilinear system $\dot{x} = Ax + ubhx$ in the plane with a scalar observer $y = Cx$:*

(1) *the pair $\{C, A\}$ is observable,*

(2) $Cb = 0$, $CAb \neq 0$.

Then the problem of exponential estimation of the phase vector with a preassigned exponent and with any bounded function $u(t)$ is solved by the observer

$$\hat{x} = \begin{pmatrix} C \\ CA \end{pmatrix}^{-1} \begin{pmatrix} y \\ \hat{\sigma} - dy \end{pmatrix}, \quad \dot{\hat{\sigma}} = \beta_1 y + \beta_2 \hat{\sigma},$$

where the parameter d satisfies condition (6.26) (*the values of the coefficients β_1 and β_2 were given above*).

Let us return to the above indicated problem of stability of the system of form (6.21), i.e.,

$$\begin{cases} \dot{\varepsilon}_1 = \varepsilon_2 - k_2\varepsilon_1 \\ \dot{\varepsilon}_2 = a(u)\varepsilon_2 - k_1\varepsilon_1, \end{cases} \tag{6.28}$$

where, for brevity, we set $a(u) = \operatorname{tr} A + a_2 u$.

We make a nondegenerate change of variables $\eta = \varepsilon_2 - k_2\varepsilon_1$, $\varepsilon = \varepsilon_1$ (and will assume that $k_2 = \text{const}$, whereas the parameter k_1 may be a function dependent on $u(t)$). Then, instead of (6.28) we obtain

$$\begin{cases} \dot{\varepsilon} = \eta \\ \dot{\eta} = -(k_2 - a(u))\eta - (k_1 - a(u)k_2)\varepsilon. \end{cases} \tag{6.29}$$

Let us consider the Lyapunov function $v = \frac{\varepsilon^2}{2} + q\frac{\eta^2}{2}$, $q = \text{const} > 0$. By virtue of the system, its derivative

$$\dot{v} = \varepsilon\eta + q\eta[-(k_2 - a(u))\eta - (k_1 - a(u)k_2)\varepsilon].$$

We set $k_1(u) = a(u)k_2 + k_1^0$, where $k_1^0 = \text{const} > 0$. In this notation

$$\dot{v} = (1 - qk_1^0)\varepsilon\eta - q\eta^2(k_2 - a(u)),$$

and, upon the fulfillment of the conditions

$$k_1^0 = 1/q, \quad k_2 > \operatorname{tr} A + |a_2|u_0 \geq \max_{|u| \leq u_0} a(u), \tag{6.30}$$

it follows that $\dot{v} \leq 0$ and, according to the Barbashin–Krasovskii theorem (since the manifold $\eta = 0$ does not contain integral trajectories), system (6.29) is asymptotically stable, and this solves the observation problem.

It is clear that upon the requisite increase of the feedback parameters k_1^0 and k_2 we can obtain any preassigned degree of stability of the observer (we omit the proof). The following theorem is valid.

Theorem 6.9. *Suppose that the following conditions are fulfilled in the bilinear system* $\dot{x} = Ax + ubhx$ *in the plane with a scalar output* $y = Cx$:

(1) *the pair* $\{C, A\}$ *is observable,*

(2) $Cb = 0$, $CAb \neq 0$.

Then an observer of the form

$$\begin{cases} \dot{\hat{y}}_1 = \hat{y}_2 - k_2(\hat{y}_1 - y) \\ \dot{\hat{y}}_2 = (ua_1 - \det A)y + (\operatorname{tr} A + a_2 u)\hat{y}_2 - k_1(\hat{y}_1 - y) \end{cases}$$

with a variable coefficient $k_1 = (\operatorname{tr} A + a_2 u)k_2 + k_1^0$ *under the requisite choice of constants* k_1^0 *and* k_2 *solves, exponentially precisely, the observation problem with any preassigned exponent and for any bounded function* $u(t)$.

Let us finally consider the case (a) from Lemma 6.5. In this most general case, the conditions

$$Cb \neq 0, \quad \det\begin{pmatrix} C \\ h \end{pmatrix} \neq 0, \quad \left|\det\begin{pmatrix} C \\ CA \end{pmatrix}\right| > u_0|Cb| \det\begin{pmatrix} C \\ h \end{pmatrix}$$

are fulfilled.

In contrast to the preceding items, now the restriction imposed on the "amplitude" of the control $u(t)$ is essential. Let us consider some possibilities of synthesizing observers, they differ by the properties of the numerator of the transfer function

$$W(s) = C(sE - A)^{-1}b = \frac{\beta(s)}{\alpha(s)}.$$

Note that $Cb \neq 0$, $\deg \beta(s) = 1$, and therefore it is relevant to consider two versions:

A. $\beta(s)$ is a Hurwitz polynomial,

B. $\beta(s)$ is a non-Hurwitz polynomial.

Version A. In this situation, for estimating the phase vector of the bilinear system

$$\dot{x} = Ax + b\tilde{u}, \quad \tilde{u} = uhx$$

with observation $y = Cx$, using a nondegenerate transformation, we reduce the system to the canonical form

$$\begin{cases} \dot{y}_1 = \lambda y_1 + y_2 \\ \dot{y}_2 = \alpha_1 y_1 + \alpha_2 y_2 + Cb\tilde{u}, \end{cases} \tag{6.31}$$

where $y = y_2$ is the output of the system and $\lambda < 0$ is a root of the numerator of the transfer function defined in Lemma 6.2 (α_1 and α_2 are constants defined by the parameters of the system).

In order to reconstruct the phase vector (y_1, y_2) and, consequently, the original vector x, it suffices to construct an observer for the first coordinate y_1 since y_2 is the known output of the system.

Under these conditions we can use an observer of a lowered order

$$\dot{\hat{y}}_1 = \lambda \hat{y}_1 + y. \tag{6.32}$$

In this case, the observation error $\varepsilon = \hat{y}_1 - y_1$ satisfies the equation

$$\dot{\varepsilon} = \lambda \varepsilon$$

which, obviously, is asymptotically stable. Thus we have the following theorem.

Theorem 6.10. *Suppose that the following conditions are fulfilled for the bilinear system $\dot{x} = Ax + ubhx$ in the plane with a scalar output $y = Cx$ and an arbitrary bounded function $|u(t)| \le u_0$:*

(1) *the pair $\{C, A\}$ is observable,*

(2) *the pair $\{A, b\}$ is controllable,*

(3) $Cb \neq 0$,

(4) $\det \binom{C}{h} \neq 0$,

(5) *the transfer function $W(s) = C(sE - A)^{-1}b$ is of minimal phase.*

Then observer (6.32) *solves the estimation problem exponentially precisely.*

Remark 6.11. It should be noted that the degree of stability of the indicated observer cannot be chosen arbitrarily, as in the observers described above, but is defined by the parameters of the system.

Version B. Suppose now that the numerator of the transfer function $W(s) = \frac{\beta(s)}{\alpha(s)}$ is unstable. Then the observer of a lowered order indicated above cannot be used.

The possibility of estimation of the phase vector of the system is extended considerably if we use in the observer nonstationary feedback dependent on the function $u(t)$. We shall show this. From the initial variables $x = \binom{x_1}{x_2}$ we pass to variables $x = \binom{\sigma}{y}$, where y is the output being measured and $\sigma = dx$, where the vector d is given in Lemma 6.2. In this case, $\det \binom{d}{C} \neq 0$ and the indicated change of coordinates is nondegenerate.

For convenience, we introduce notation

$$Cbh \begin{pmatrix} d \\ C \end{pmatrix}^{-1} = (\tilde{a}_1, \tilde{a}_2), \quad CA \begin{pmatrix} d \\ C \end{pmatrix}^{-1} = (\hat{a}_1, \hat{a}_2).$$

In this notation and with due account of the properties of the vector d indicated in Lemma 6.2, the equations of the system being considered, in the new coordinates, assume the form

$$\begin{cases} \dot{\sigma} = \lambda\sigma - y \\ \dot{y} = (\hat{a}_1 + u\tilde{a}_1)\sigma + (\hat{a}_2 + u\tilde{a}_2)y, \end{cases} \tag{6.33}$$

where λ is a number from Lemma 6.2.

To estimate the phase vector, we use a full-dimensional observer

$$\begin{cases} \dot{\hat{\sigma}} = \lambda\hat{\sigma} - y - k_2(\hat{y} - y) \\ \dot{\hat{y}} = (\hat{a}_1 + u\tilde{a}_1)\hat{\sigma} + (\hat{a}_2 + u\tilde{a}_2)y - k_1(\hat{y} - y), \end{cases} \tag{6.34}$$

where the choice of feedback parameters of the observer k_1 and k_2 will serve for ensuring the exponential stability of the system in deviations

$$\begin{cases} \dot{\varepsilon}_1 = \lambda\varepsilon_1 - k_2\varepsilon_2 \\ \dot{\varepsilon}_2 = a(u)\varepsilon_1 - k_1\varepsilon_2, \end{cases} \tag{6.35}$$

where $a(u) = \hat{a}_1 + u\tilde{a}_1$, $\varepsilon_1 = \hat{\sigma} - \sigma$, and $\varepsilon_2 = \hat{y} - y$ is the known output of the system.

We shall show that such a choice of parameters k_1 and k_2 is possible, moreover, it can ensure any preassigned degree of the exponential stability of system (6.35).

Note that under the conditions of strict uniform observability the function $a(u)$ is alternating and, moreover,

$$0 < a_* \le |a(u)| \le a^*. \tag{6.36}$$

Indeed, the condition of strict, uniform with respect to t, observability (6.13) is invariant relative to a change of variables, and therefore, by virtue of system (6.33), where

$$A = \begin{pmatrix} \lambda & -1 \\ \hat{a}_1 & \hat{a}_2 \end{pmatrix}; \quad b = \begin{pmatrix} 0 \\ 1 \end{pmatrix}; \quad \begin{pmatrix} h = (\tilde{a}_1, \tilde{a}_2) \\ C = (0, 1) \end{pmatrix},$$

assumes the form

$$u_0 < \left| \frac{\det\begin{pmatrix} C \\ CA \end{pmatrix}}{Cb \det\begin{pmatrix} C \\ h \end{pmatrix}} \right| = \frac{|\hat{a}_1|}{|\tilde{a}_1|},$$

whence it follows immediately that the function $a(u) = \tilde{a}_1 u + \hat{a}_1$ is of constant sign, $a_* = |\hat{a}_1| - |\tilde{a}_1|u_0$, $a^* = |\hat{a}_1| + |\tilde{a}_1|u_0$.

Obviously, for solving an observation problem it is sufficient that $\varepsilon_1 \to 0$. For $k_2 =$ const this variable satisfies the equation

$$\ddot{\varepsilon}_1 + \alpha_2(u)\dot{\varepsilon}_1 + \alpha_1(u)\varepsilon_1 = 0, \tag{6.37}$$

where $\alpha_2(u) = k_1 - \lambda$, and $\alpha_1(u) = k_2 a(u) - \lambda k_1$. In order to analyze the stability of this equation, we use the Lyapunov function

$$V = \varepsilon^T P\varepsilon, \quad \text{where} \quad \varepsilon^T = (\varepsilon_1, \dot{\varepsilon}_1), \quad P = \begin{pmatrix} \phi & \gamma \\ \gamma & \psi \end{pmatrix} > 0.$$

By virtue of the system the derivative of the function V has the form $\dot{V} = \varepsilon^T D(u)\varepsilon$, where

$$D(u) = \begin{pmatrix} -2\alpha_1\gamma & \phi - \lambda_2\gamma - \lambda_1\psi \\ \phi - \lambda_2\gamma - \lambda_1\psi & 2(\gamma - \phi\alpha_2) \end{pmatrix}.$$

For equation (6.37) to be exponentially stable with the defined exponent $2\eta > 0$, it is sufficient that, for all $t \geq 0$, the matrix inequality

$$\max_{t>0}(D(u) + 2\eta P) < 0 \tag{6.38}$$

should be fulfilled which, by virtue of the Silvester criterion and the condition of strict uniform observability, is equivalent to the inequalities

$$\max_{|u|\leq u_0}(-2\alpha_1\gamma + 2\eta\phi) < 0, \qquad \min_{|u|\leq u_0} \det(D(u) + 2\eta P) > 0. \tag{6.39}$$

In the sequel we assume that $\gamma > 0$, and then from (6.39) we have

$$\alpha_1(u) > \frac{\eta\phi}{\gamma} = w_1 > 0. \tag{6.40}$$

Since [8]

$$\det(D(u) + 2\eta P) = 4\eta^2 \det P + 2\eta(\operatorname{tr} D \operatorname{tr} P - \operatorname{tr} PD) + \det D \tag{6.41}$$

and since we can easily make sure by a direct verification that $\det D = 4\alpha_1 \det P - (\phi + \alpha_1\psi - \alpha_2\gamma)^2$ and $\operatorname{tr} D \operatorname{tr} P - \operatorname{tr} PD = -2\alpha_2 \det P$, it follows that the solution of the posed problem is given by the following choice: $k_2 = k_2^0 \cdot \operatorname{sgn}(a(u)) = \text{const}$ (since the function $a(u)$) is of constant sign), $k_2^0 > 0$, and $k_1 = k_1(u)$ such that the relation $\phi + \alpha_1\psi - \alpha_2\gamma = 0$ is satisfied for all $u(t)$, i.e.

$$k_1(u) = \frac{\phi + \psi k_2^0|a(u)| + \lambda\gamma}{\psi\lambda + \gamma}, \qquad \psi\lambda + \gamma \neq 0. \tag{6.42}$$

In this case, $\alpha_2(u) = (\phi + \alpha_1(u))/\gamma$. With due account of the last relation and the fact that $\det P > 0$, condition (6.41) becomes an inequality

$$\eta^2 - \alpha_2(u)\eta + \alpha_1(u) = \frac{1}{\gamma}\big(\alpha_1(u)(\gamma - \psi\eta) - (\phi\eta - \gamma\eta^2)\big) > 0,$$

which, for $\gamma > \psi\eta > 0$, is equivalent to the inequality

$$\alpha_1(u) > \gamma\frac{\phi\eta - \gamma\eta^2}{\gamma - \psi\eta} = w_2. \tag{6.43}$$

Thus, for $k_1(u)$ from (6.42) for an exponential stability of system (6.37) with exponent 2η it is sufficient that the estimate

$$\alpha_1(u) = k_2^0|a(u)| - \lambda k_1 > \max\{w_1, w_2\} = w$$

be valid. This condition holds if

$$k_2^0|a(u)|\frac{\gamma}{\lambda\psi+\gamma} > w + \frac{\lambda\phi+\lambda^2\gamma}{\lambda\psi+\gamma} = \tilde{w} = \text{const}\,.$$

Let $\psi = 1$, $\gamma > \max\{\eta, \lambda\}$, $\phi > \gamma^2$, and then the posed problem is solved by the choice of $k_1(u)$ from (6.42) and

$$k_2^0 > \frac{\tilde{w}(\lambda+\gamma)}{a_*\gamma}. \tag{6.44}$$

We have thus proved the following theorem.

Theorem 6.12. *Suppose that the following conditions are fulfilled for the bilinear system $\dot{x} = Ax + ubhx$ in the plane with a scalar output $y = Cx$:*

– the triple $\{C, A\}$ is in the general position,

– condition (6.13) *of strict, uniform with respect to t, observability is fulfilled.*

Then there exist a number k_2 and a function $k_1(u)$ such that the observer

$$\begin{cases} \dot{\hat{\sigma}} = \lambda\hat{\sigma} - y - k_2(\hat{y} - y) \\ \dot{\hat{y}} = (\hat{a}_1 + u\tilde{a}_1)\hat{\sigma} + (\hat{a}_2 + u\tilde{a}_2)y - k_1(\hat{y} - y) \end{cases}$$

solves exponentially precisely the observation problem with any preassigned exponent of accuracy for any uniformly bounded function $u(t)$: $|u(t)| \le u_0$.

Remark 6.13. It stands to reason that the observer indicated in Theorem 6.12 is also suitable for minimal-phase systems.

6.2 Asymptotic observers for certain classes of n-dimensional bilinear systems

6.2.1 Problem statement

We consider a problem of construction of asymptotic observer for the bilinear controllable system with a linear output

$$\begin{cases} \dot{x} = Ax + uBx \\ y = Cx, \end{cases} \tag{6.45}$$

where $x \in \mathbb{R}^n$, $A, B \in \mathbb{R}^{n\times n}$, $C \in \mathbb{R}^{l\times n}$ are known parameters of the system, $u(t)$ is a scalar input, and $y(t) \in \mathbb{R}^l$ is a measurable l-dimensional output of the system. We have to construct an estimate $\tilde{x}(t)$ which asymptotically (exponentially) converges to the unknown phase vector $x(t)$.

For bilinear systems not only the construction of the observer itself is difficult but also obtaining the conditions of observability of system (6.45). In particular, if we regard (6.45) as a nonlinear, affine with respect to u, system, then, in accordance with [84], the condition of reconstructibility has the form

$$\begin{gathered} \dim(\operatorname{conv}(C; CQ_1; CQ_1Q_2; CQ_1Q_2Q_3, \dots)) = n, \\ Q_i = A \quad \text{or} \quad Q_i = B, \end{gathered} \tag{6.46}$$

where conv$(\cdot)$ is a convex hull of the indicated set of vectors. Condition (6.46) means that any initial state $x(0)$ is reconstructible with respect to the measurements of the output $y(t)$ for a suitable input $u(t)$. However, when observers are constructed the function $u(t)$ is, as a rule, defined and cannot be changed for solving the observation problem. We can easily indicate a situation where condition (6.46) is fulfilled but the system is nonobservable, for instance, when the pair $\{C, A\}$ is observable

$$\dim(\operatorname{conv}(C; CA; CA^2, \dots, CA^{n-1})) = n$$

and condition (6.46) is fulfilled for any matrix B, but for $B = A$ and $u = -1$, system (6.45) assumes the form

$$\dot{x} = 0, \quad y = Cx(0)$$

and, consequently, we cannot reconstruct $x(0)$ using the measurements of $y(t)$.

Thus, conditions (6.46) are not uniform relative to $u(t)$. If the function $u(t)$ is known, then system (6.45) can be regarded as a linear system with a nonstationary matrix $\tilde{A}(t) = A + u(t)B$:

$$\begin{cases} \dot{x} = \tilde{A}(t)x \\ y = Cx. \end{cases}$$

If the function $u(t)$ is differentiable a sufficient number of times, then the observability condition assumes the form

$$\begin{gathered} \dim(Q_1(t), Q_2(t), \dots, Q_n(t)) = n, \quad t \geq 0, \\ Q_1 = C; \quad Q_{i+1}(t) = \frac{dQ_i(t)}{dt} + Q_i(t)\tilde{A}(t). \end{gathered} \tag{6.47}$$

If the function $u(t)$ does not possess the required smoothness or is unknown, the indicated reduction to a problem of linear observation is impossible.

The aim of this section is to obtain sufficient conditions of uniform, with respect to $u(t)$, observability of system (6.45) and to construct asymptotic observers under these conditions. Generally speaking, the independence of the solution of the problem of the particular input $u(t)$ is possible only for the generate matrix B, and therefore the specific character of the observers proposed in the sequel is defined by different forms of degeneration of the matrix B, and this reduces the problem under consideration to an observation problem for a linear system with an unknown input.

6.2.2 Systems with a scalar output and a degenerate matrix of bilinearity

Suppose that in system (6.45) with a scalar output (i.e., $l = 1$) the matrix of bilinearity B is of a minimal rank, i.e., rank $B = 1$. Then $B = bh$, where b and h are known vector-column and vector-row, respectively.

In this case, we can easily obtain sufficient conditions of uniform, with respect to $u(t)$, observability of the system. Indeed, if $Cb = 0$, $CAb = 0, \ldots, CA^{n-2}b = 0$, then $Q_i(t)$ in (6.47) has the form

$$Q_1 = C, Q_2 = CA, \ldots, Q_n = CA^{n-1},$$

and, consequently, all Q_i are independent of $u(t)$ and system (6.45) is uniformly, with respect to $u(t)$, observable if

$$\dim(C, CA, \ldots, CA^{n-1}) = n,$$

i.e., if the pair $\{C, A\}$ is observable. Then the following theorem is valid.

Theorem 6.14. *The fulfillment of the following conditions is sufficient for the uniform, with respect to $u(t)$, observability of system* (6.45):

(1) $B = bh$, $b \in \mathbb{R}^{n\times 1}$, $h \in \mathbb{R}^{1\times n}$,

(2) $Cb = 0$, $CAb = 0, \ldots, CA^{n-2}b = 0$,

(3) *the pair* $\{C, A\}$ *is observable.*

Here is the technique of constructing an observer for systems of this kind. We can write the bilinear system (6.45) in the form which is standard for linear systems:

$$\begin{cases} \dot{x} = Ax + b\bar{u} \\ y = Cx, \end{cases} \tag{6.48}$$

where $\bar{u} = uhx$ is an unknown scalar input signal. Suppose, in addition, that the pair $\{A, b\}$ is controllable. Then the transfer function

$$W(s) = C(sI - A)^{-1}b = \frac{\beta_m(s)}{\alpha_n(s)} \tag{6.49}$$

is defined for system (6.48). Here $\beta_m(s)$ and $\alpha_n(s)$ are coprime polynomials of s of orders m and n, respectively ($0 \le m < n$).

Under condition (2) from Theorem 6.14 the relative order of the system $r = n - m$ is maximal, i.e., $r = n$, $m = 0$. Then, using a nondegenerate change of coordinates with matrix P, we can reduce system (6.45) to the canonical form with the isolation of zero dynamics (whose dimension in this case is zero):

$$\begin{cases} \dot{x}_1 = x_2 \\ \vdots \\ \dot{x}_{n-1} = x_n \\ \dot{x}_n = -a_1 x_1 - \cdots - a_n x_n + u\tilde{h}x, \end{cases} \tag{6.50}$$

where $\tilde{h} = hP^{-1}$, $\alpha_n(s) = \det(sI - A) = s^n + a_n s^{n-1} + \cdots + a_1$, $y = x_1$ (here, for simplicity, we preserve the old notation for the phase vector). If the input $u(t)$ is known and uniformly bounded, i.e., $|u(t)| \le u_0$ for $t \ge 0$, then, for constructing an estimate of the vector $x(t)$, we use the observer

$$\begin{cases} \dot{\tilde{x}}_1 = \tilde{x}_2 - k_1(\tilde{x}_1 - y) \\ \vdots \\ \dot{\tilde{x}}_{n-1} = \tilde{x}_n - k_{n-1}(\tilde{x}_1 - y) \\ \dot{\tilde{x}}_n = -a_1\tilde{x}_1 - \cdots - a_n\tilde{x}_n + u\tilde{h}\tilde{x} - k_n(\tilde{x}_1 - y). \end{cases} \tag{6.51}$$

Then the solution of the observation problem reduces to the provision of stability of the system in deviations $e = \tilde{x} - x$ described by the equation

$$\dot{e} = \tilde{A}_k e + \tilde{b}(u\tilde{h}e - ae), \tag{6.52}$$

where

$$\tilde{A}_k = \begin{pmatrix} -k_1 & 1 & 0 & \dots & 0 \\ -k_2 & 0 & 1 & \dots & 0 \\ \vdots & \vdots & \vdots & \vdots & \\ -k_n & 0 & 0 & \dots & 0 \end{pmatrix}, \quad \tilde{b} = \begin{pmatrix} 0 \\ \vdots \\ 0 \\ 1 \end{pmatrix}, \quad a = (a_1, \dots, a_n).$$

The characteristic polynomial of the matrix $\tilde{A}_k$ has the form $\psi(s) = \det(sI - \tilde{A}_k) = s^n + k_1 s^{n-1} + \cdots + k_n$ and is defined by a requisite choice of parameters $k_1, \dots, k_n$. If we set

$$k_i = K^i q_i, \quad K = \text{const} > 0, \tag{6.53}$$

where K is the amplification factor and q_i are the coefficients of the arbitrarily defined Hurwitz polynomial $\psi_0(s) = s^n + q_1 s^{n-1} + q_n = (s + \lambda_1) \cdots (s + \lambda_n)$ whose roots $(-\lambda_i)$ satisfy the relations $\lambda_1 = 1$, $\lambda_{i+1} > \lambda_i$, $i = 1, \dots, n-1$, then $(-K\lambda_i)$ are roots of the polynomial $\psi(s)$. We have the following theorem.

Theorem 6.15. *Suppose that the following conditions are fulfilled for system* (6.45) *for* $C \in \mathbb{R}^{1\times n}$:

(1) rank $B = 1$, $B = hb$,
(2) *the pair* $\{A, b\}$ *is controllable and the pair* $\{C, A\}$ *is observable,*
(3) $Cb = 0$, $CAb = 0$, $\dots$, $CA^{n-2}b = 0$, $CA^{n-1}b \neq 0$,
(4) *the control* $u(t)$ *is bounded, i.e.,* $|u(t)| \le u_0$ *for* $t \ge 0$.

Then observer (6.51) *with coefficients* (6.53) *for* $K > K_0$*, where* $K_0 =$ const, *depends only on the parameters of the original system* (6.45) *and on* u_0*, exponentially reconstructs the phase vector of the system. For the estimation error* $e(t) = \tilde{x}(t) - x(t)$ *there is an estimate* $|e(t)| \le Q|e(0)|e^{-(K-K_0)t}$*, where* $Q =$ const, *depends on the parameters of the system and on* K *(generally speaking,* $Q \to +\infty$ *as* $K \to +\infty$*).*

Proof. The expansion

$$e^{\tilde{A}_k t} = \sum_{i=0}^{n-1} \alpha_i(t)\tilde{A}_k^i$$

holds for the matrix exponential $e^{\tilde{A}_k t}$.

As was shown in Chap. 5, for the choice of coefficients k_i indicated above, we have estimates

$$|\alpha_i(t)| \le \frac{N_i}{K^i} e^{-Kt}, \tag{6.54}$$

where $N_i =$ const > 0 do not depend on the amplification factor K. In this case, the estimate

$$\begin{aligned} |e(t)| &\le |e^{\tilde{A}_k t} e(0)| + \left| \int_0^t e^{\tilde{A}_k(t-\tau)} \tilde{b}\tilde{u}(\tau)\,d\tau \right| \\ &\le Q|e(0)|e^{-Kt} + \int_0^t \left|e^{\tilde{A}_k \tau}\tilde{b}\right| |\tilde{u}(t-\tau)|\,d\tau \\ &= Q|e(0)|e^{-Kt} + \int_0^t \sum_{i=0}^{n-1} |\alpha_i(\tau)| \left|\tilde{A}_k^i \tilde{b}\right| |\tilde{u}(t-\tau)|\,d\tau, \end{aligned}$$

where $Q =$ const > 0 and $\tilde{u} = u\tilde{h}e - ae$, is valid for the phase vector $e(t)$ of the system in deviations (6.52). We can make sure by direct verification that $Q_i = |\tilde{A}_k^i \tilde{b}| > 0$ ($i = 0, \dots, n-1$) does not depend on the choice of the amplification factor K either.

Without loss of generality, we set $K > 1$ and assume that the input function $u(t)$ is bounded, i.e., $|u(t)| \le u_0$. Then

$$\begin{aligned} |e(t)| &\le Q|e(0)|e^{-Kt} + \left(\sum_{i=0}^{n-1} N_i Q_i \right) (u_0|\tilde{h}| + |a|) \int_0^t e^{-K\tau}|e(t-\tau)|\,d\tau \\ &= Q'e^{-Kt} + K_0 \int_0^t e^{-K(t-\tau)}|e(\tau)|\,d\tau, \end{aligned}$$

where K_0 does not depend on K and $Q' = Q|e(0)|$. We multiply both sides of the last inequality by e^{Kt} and obtain

$$|e(t)|e^{Kt} \leq Q' + K_0 \int_0^t |e(\tau)|e^{K\tau}\, d\tau,$$

whence, by virtue of the Gronwall–Bellman lemma,

$$|e(t)|e^{Kt} \leq Q' + Q'(e^{K_0 t} - 1) = Q' e^{K_0 t}.$$

The final estimate has the form

$$|e(t)| \leq Q' e^{-(K-K_0)t},$$

whence it follows that for $K > K_0$ observer (6.51) solves the problem exponentially precisely with any preassigned rate of convergence. Note, however, that the magnitude of the constant $Q' > 0$ depends on K (in general, Q' grows with the increase of K). The theorem is proved. □

If in system (6.45) the input $u(t)$ is unknown but bounded, then this approach is suitable only in asymptotics as $K \to \infty$.

The proposed approach can be generalized to bilinear systems with a degenerate matrix of bilinearity with an arbitrary relative order. Suppose, for instance, that the relative order of system (6.48) is equal to $r < n$, i.e.,

$$Cb = 0, \quad CAb = 0, \quad \dots, \quad CA^{r-2}b = 0, \quad CA^{r-1}b \neq 0,$$

and the numerator of the transfer function $W(s)$, which is a polynomial $\beta_m(s)$, is a Hurwitz polynomial (in this case, $m = n - r$). Then system (6.48) can be reduced, by a nondegenerate transformation, to the canonical form with isolation of zero dynamics

$$\left\{\begin{array}{l} \dot{x}' = A_{11}x' + A_{12}y \\ \left\{\begin{array}{l} \dot{y}_1 = y_2 \\ \quad\vdots \\ \dot{y}_{r-1} = y_r \\ \dot{y}_r = -a'x' - a''\bar{y} + u(b'x' + b''\bar{y}), \quad y = y_1, \end{array}\right. \end{array}\right. \tag{6.55}$$

where $x' \in \mathbb{R}^{n-r}$, $\bar{y} = (y_1, \dots, y_r)^\top$, and the vectors a', a'', b' and b'' of the corresponding dimensions are defined by the parameters of the original system, and $A_{12} = (0, \dots, 0, 1)^\top$. Moreover, $\det(sI - A_{11}) = \beta_m(s)$, and we assume in what follows that A_{11} is a Hurwitz matrix.

The required exponential estimate of the phase vector $\binom{x'}{\bar{y}}$ is given by the observer

$$\begin{cases} \dot{\tilde{x}}' = A_{11}\tilde{x}' + A_{12}y \\ \begin{cases} \dot{\tilde{y}}_1 = \tilde{y}_2 - k_1(\tilde{y}_1 - y) \\ \vdots \\ \dot{\tilde{y}}_{r-1} = \tilde{y}_r - k_{r-1}(\tilde{y}_1 - y) \\ \dot{\tilde{y}}_r = -a'\tilde{x}' - a''\tilde{y} + u(b'\tilde{x}' + b''\tilde{y}) - k_r(\tilde{y}_1 - y), \end{cases} \end{cases} \tag{6.56}$$

where $\tilde{y} = (\tilde{y}_1, \dots, \tilde{y}_r)^\top$. Its efficiency is based on the stability of the system in deviations

$$\begin{cases} \dot{\varepsilon}' = A_{11}\varepsilon' \\ \begin{cases} \dot{e}_1 = e_2 - k_1 e_1 \\ \vdots \\ \dot{e}_{r-1} = e_r - k_{r-1}e_1 \\ \dot{e}_r = -k_r e_1 - (a'' + ub'')e - (a' + ub')\varepsilon', \end{cases} \end{cases} \tag{6.57}$$

where $\varepsilon' = \tilde{x}' - x' \in \mathbb{R}^{n-r}$, $e = \tilde{y} - \bar{y} \in \mathbb{R}^r$. As before, we choose a stable polynomial $\psi_0(s) = (s + \lambda_1) \cdots (s + \lambda_r) = s^r + q_1 s^{r-1} + \cdots + q_r$ with roots $(-\lambda_i)$, where $\lambda_1 = 1$, $\lambda_{i+1} > \lambda_i$ for $i = 1, \dots, r-1$ and set the coefficients of the observer k_i in the form $k_i = K^i q_i$, $K = \text{const} > 0$. Then we have the following theorem.

Theorem 6.16. *Suppose that conditions* (1), (2), *and* (4) *from Theorem 6.15 are fulfilled for system* (6.45), *the relative order of the system is equal to* r, *and* $\beta_m(s)$ *is a Hurwitz polynomial. Then there exists a constant* K_0, *dependent on the parameters of the system and on* u_0, *such that for* $K > K_0$ *observer* (6.56) *reconstructs exponentially the phase vector of system* (6.45).

The *proof* of Theorem 6.16 is similar to that of Theorem 6.15 with the only difference that the estimate now contains an additional exponentially decreasing term connected with the estimate ε'

Remark 6.17. If the relative order $r < n$, then the degree of stability of the system in deviations (6.57) does not exceed the degree of stability of the polynomial $\beta_m(s)$, in contrast to the case $r = n$.

Remark 6.18. Without essential changes we can generalize the described method of estimation of the phase vector to systems with a degenerate matrix of bilinearity of the form $B = bh + dC$. In this case, the problem is solved by observer of form (6.51) with an additional term dy on the right-hand side, i.e.,

$$\dot{\tilde{x}} = A\tilde{x} - \bar{k}(C\tilde{x} - y) + dy.$$

6.2.3 Systems with a vector output and degenerate matrix of bilinearity

Additional possibilities for synthesizing observers for bilinear systems appear in the case of vector output, i.e., for $l > 1$, $C \in \mathbb{R}^{l\times n}$, $\operatorname{rank} C = l$.

Let us consider the degeneration of a matrix of bilinearity of the form $B = \bar{B}\bar{H}$, where $\bar{B} \in \mathbb{R}^{n\times m}$, $\bar{H} \in \mathbb{R}^{m\times n}$, under the condition $m < l$. Under these assumptions system (6.45) can again be written as a linear system with an unknown input. As in (6.48), we have

$$\begin{cases} \dot{x} = Ax + \bar{B}\bar{u} \\ y = Cx, \end{cases} \tag{6.48'}$$

where $\bar{u} = u\bar{H}x$ is an unknown input signal. The methods of synthesis of observers of these systems are based on the algorithms of synthesis of observers for linear systems with uncertainty which are described in detail in Chap. 5. We shall only indicate the main results.

For constructing an exponential observer for system (6.48′) it is sufficient that the following conditions should be fulfilled:

(1°) $\operatorname{rank} C = l$, $\operatorname{rank} \bar{B} = m$, $m < l$,

(2°) $\operatorname{rank} C\bar{B} = m$, $C\bar{B} \in \mathbb{R}^{l\times m}$,

(3°) the invariant zeros of system (6.48′) are absent or stable.

Under these conditions, system (6.48′) can be reduced, by means of nondegenerate change of coordinates, to the form

$$\begin{cases} \dot{x}' = A_{11}x' + A_{12}y' \\ \dot{y}' = A_{21}x' + A_{22}y' + B'\bar{u}, \end{cases} \tag{6.58}$$

where y' are m coordinates from the output vector y, $x' \in \mathbb{R}^{n-m}$ is the remaining unknown part of the phase vector, A_{ij}, B' are matrices with constant coefficients of corresponding dimensions, $\det B' \neq 0$.

Using the remaining $(l - m)$ components of the output y, we define a new output $\tilde{y} = \tilde{C}x'$ for system (6.58). For reconstructing the unknown part of the phase vector x' we can use an observer of the form

$$\dot{\tilde{x}}' = A_{11}\tilde{x}' + A_{12}y' - L(\tilde{C}\tilde{x}' - \tilde{y}), \tag{6.59}$$

where the matrix $L \in \mathbb{R}^{(n-m)\times(n-l)}$ is chosen from the condition that $A_L = A_{11} - L\tilde{C}$ is a Hurwitz matrix. It was shown in Chap. 5 that if system (6.48′) does not have invariant zeros, then the pair $\{\tilde{C}, A_{11}\}$ is observable, and if stable invariant zeros are present, then this pair is reconstructible. Therefore, if condition (3°) is fulfilled, the indicated matrix L exists.

The following theorem is valid.

Theorem 6.19. *Suppose that for system* (6.45) *the matrix of bilinearity B has a degeneration of the form $B = \bar{B}\bar{H}$, $\bar{B} \in \mathbb{R}^{n\times m}$. Suppose, in addition, that conditions*

(1°)–(3°) *are fulfilled. Then observer* (6.59) *gives an exponential estimate of the unknown part of the phase vector.*

Remark 6.20. It should be emphasized that in contrast to a scalar output, we do not use the information about the input $u(t)$ when constructing the described observer.

Remark 6.21. The proposed approach can be generalized to the following classes of bilinear systems.

1°. Systems with bilinearity matrix of the form $B = \bar{B}\bar{H} + DC$.

2°. Systems in which in addition to a bilinear component there is a linear component, i.e., systems of the form

$$\begin{cases} \dot{x} = Ax + uBx + Du' \\ y = Cx, \end{cases} \tag{6.45'}$$

where $D \in \mathbb{R}^{n\times q}$. If the function $u'(t)$ is known, then, if we use the model

$$\begin{cases} \dot{\tilde{x}} = A\tilde{x} + uB\tilde{x} + Du' \\ \tilde{y} = C\tilde{x}, \end{cases}$$

then the problem reduces to the construction of an observer for the system in deviations $e = \tilde{x} - x$

$$\begin{cases} \dot{e} = Ae + uBe, \\ \varepsilon = Ce \end{cases}$$

which is identical to system (6.45).

3°. By analogy we can solve the problem of constructing an observer for a bilinear system with a k-dimensional input in the case of the degeneration of the bilinearity matrices. We shall explain this using the example of a system

$$\begin{cases} \dot{x} = Ax + \sum_{i=1}^{k} u_i B_i x \\ y = Cx \end{cases} \tag{6.45''}$$

in which all bilinearity matrices B_i are of a minimal rank (the case of degeneration of bilinearity matrices with arbitrary ranks can be considered according to the scheme described above)

$$B_i = b_i h_i, \quad b_i \in \mathbb{R}^{n\times 1}, \quad h_i \in \mathbb{R}^{1\times n}.$$

We set

$$\bar{B} = (b_1, \dots, b_k), \quad \bar{U} = \begin{pmatrix} u_1 h_1 x \\ \vdots \\ u_k h_k x \end{pmatrix}$$

and reduce system (6.45″) to the standard form (6.48′)

$$\begin{cases} \dot{x} = Ax + \bar{B}\bar{u} \\ y = Cx. \end{cases}$$

The observer of the system is constructed according to the scheme described above.

6.2.4 Systems with vector output and known input

Let us consider system (6.45) with an arbitrary (not necessarily degenerate) matrix of bilinearity. For a sufficiently large l, for this system to be observable for a given $u(t)$, it may be sufficient that the condition $\dim(Q_1(t), Q_2(t)) = n$ is satisfied. Since $Q_1(t) = C$ and $Q_2(t) = CA + CBu(t)$, the sufficient condition of observability assumes the form

$$\operatorname{rank}\begin{pmatrix} C \\ CA + CBu(t) \end{pmatrix} = n, \quad t \geq 0. \tag{6.60}$$

Note that in this condition the continuity or differentiability of the function $u(t)$ is not required. Since $C \in \mathbb{R}^{l \times n}$, condition (6.60) can be fulfilled only for $l \geq n/2$. Let us consider system (6.45) under this condition.

Since $\operatorname{rank} C = l$, it follows that, without loss of generality, we can assume that

$$C = (I_{l \times l}, 0_{l \times r}).$$

Consider a nondegenerate change of coordinates of system (6.45)

$$\begin{pmatrix} y \\ z \end{pmatrix} = \begin{pmatrix} C \\ H \end{pmatrix} x,$$

where the matrix $H \in \mathbb{R}^{r \times n}$ $(r = n - l)$ is chosen from the condition $\det\left(\begin{smallmatrix} C \\ H \end{smallmatrix}\right) \neq 0$. For this purpose, it suffices, for instance, to choose $H = (\tilde{H}, I_{r \times r})$ for any $\tilde{H} \in \mathbb{R}^{r \times l}$. It is clear that for solving the original problem it suffices to obtain an estimate of the vector $z \in \mathbb{R}^r$.

Let us write the matrices A and B of the original system (6.45) in block forms

$$A = \begin{pmatrix} A_1 & A_2 \\ A_3 & A_4 \end{pmatrix}, \quad B = \begin{pmatrix} B_1 & B_2 \\ B_3 & B_4 \end{pmatrix},$$

where $A_1, B_1 \in \mathbb{R}^{l \times l}$ and $A_4, B_4 \in \mathbb{R}^{r \times r}$. Then the equation for the component z assumes the form

$$\dot{z} = (P_1 + uP_2)y + (R_1 + uR_2)z,$$

where $P_1 = \tilde{H}A_1 + A_3 - \tilde{H}A_2\tilde{H} - A_4\tilde{H}$, $P_2 = \tilde{H}B_1 + B_3 - \tilde{H}B_2\tilde{H} - B_4\tilde{H}$, $R_1 = \tilde{H}A_2 + A_4$, $R_2 = \tilde{H}B_2 + B_4$. For the known $u(t)$ we construct an observer for z in the form

$$\dot{\tilde{z}} = (P_1 + uP_2)y + (R_1 + uR_2)\tilde{z}. \tag{6.61}$$

In this case, the observation error $e = \tilde{z} - z$ satisfies the equation

$$\dot{e} = (R_1 + uR_2)e,$$

and if $R_2 = 0$ and R_1 is a Hurwitz matrix, then $e \to 0$ exponentially. Let us now choose a matrix $\tilde{H}$ from the conditions

$$\begin{cases} R_2 = \tilde{H} B_2 + B_4 = 0 \\ R_1 = \tilde{H} A_2 + A_4 = A', \end{cases} \tag{6.62}$$

where A' is an arbitrary Hurwitz matrix with a defined degree of stability.

A solution of equations (6.62) for given A_4 B_4, and A' exists [4] if

$$\operatorname{rank}(B_2, A_2) = \operatorname{rank} \begin{pmatrix} B_2 & A_2 \\ -B_4 & -A_4 + A' \end{pmatrix}. \tag{6.63}$$

Thus we have the following theorem.

Theorem 6.22. *Suppose that condition* (6.63) *is fulfilled for system* (6.45) *for a certain Hurwitz matrix A'. Then there exists a matrix $\tilde{H} \in \mathbb{R}^{r \times l}$ such that observer* (6.61) *reconstructs the phase vector of the system exponentially for any known input function $u(t)$.*

Remark 6.23. If $l \geq r$, i.e., $l \geq n/2$ and rank $B_2 = r$, then there exists $\tilde{H}$ satisfying the first equation (6.62). For such an $\tilde{H}$ the equation for the observation error does not depend on the control $u(t)$ but the stability of the matrix $R_1 = \tilde{H} A_2 + A_4$ is not guaranteed and is defined by the parameters of the system and by the degree of arbitrariness in the choice of solution $\tilde{H}$.

Remark 6.24. For condition (6.63) to be fulfilled for any A_4, B_4 and any preassigned matrix A' (which defines the asymptotics of observer (6.61)), it is sufficient that the condition

$$\operatorname{rank}(B_2, A_2) = 2r$$

be fulfilled. Taking into account that $(B_2, A_2) \in \mathbb{R}^{l \times 2r}$, for this condition to be fulfilled it is necessary that the relation $l \geq 2r = 2(n - l)$ be satisfied, i.e.,

$$l \geq \frac{2}{3} n.$$

Remark 6.25. In the case under consideration, we have

$$\begin{pmatrix} C \\ CA + CBu(t) \end{pmatrix} = \begin{pmatrix} I_l & 0 \\ A_1 + uB_1 & A_2 + uB_2 \end{pmatrix}$$

and the sufficient condition of observability assumes the form

$$\operatorname{rank}(A_2 + uB_2) = r = n - l.$$

The following sufficient condition of uniform observability holds and the following theorem is valid.

Theorem 6.26. *Let* $\operatorname{rank} C = l$, $l \geq 2(n-l)$ *(i.e.,* $l \geq \frac{2}{3}n$*), and* $\operatorname{rank}(A_2, B_2) = 2(n-l)$*. Then*

$$\operatorname{rank}\begin{pmatrix} C \\ CA + CBu(t) \end{pmatrix} = n$$

for all $u(t)$*, i.e., system* (6.45) *is observable uniformly with respect to* $u(t)$*.*

6.2.5 Asymptotic observers on the basis of the decomposition method

As in Sec. 6.2.4, we shall consider system (6.45) with an arbitrary matrix of bilinearity. Assuming that $\operatorname{rank} C = l$ ($C \in \mathbb{R}^{l\times n}$), we pass to coordinates

$$\begin{pmatrix} y \\ z \end{pmatrix} = \begin{pmatrix} C \\ H \end{pmatrix} x,$$

where the matrix $H \in \mathbb{R}^{(n-l)\times n}$ is chosen from the condition of nondegeneracy of the indicated transition.

As in Sec. 6.2.4, after the change of coordinates, the system can be written in block form

$$\begin{cases} \dot{y} = A_1 y + A_2 z + u(B_1 y + B_2 z) \\ \dot{z} = A_3 y + A_4 z + u(B_3 y + B_4 z). \end{cases} \tag{6.64}$$

Since the output $y(t)$ of the system is known, we can regard this system as a linear system with the unknown input $f = (uz) \in \mathbb{R}^{n-l}$ and the known input $u' = (uy) \in \mathbb{R}^l$, i.e.,

$$\begin{pmatrix} \dot{y} \\ \dot{z} \end{pmatrix} = A \begin{pmatrix} y \\ z \end{pmatrix} + B'u' + B''f, \tag{6.65}$$

where

$$A = \begin{pmatrix} A_1 & A_2 \\ A_3 & A_4 \end{pmatrix}, \quad B' = \begin{pmatrix} B_1 \\ B_3 \end{pmatrix}, \quad B'' = \begin{pmatrix} B_2 \\ B_4 \end{pmatrix}. \tag{6.66}$$

Let us consider the case where $l > n - l$, i.e., $l > \frac{n}{2}$. Under this condition, system (6.65) can be regarded as a linear stationary hyperoutput system with an input $f(t)$ without certainty, and, consequently, for solving the observation problem we can use the methods of synthesis of observers for hyperoutput systems which were described in detail in Chap. 5.

Here are the main results.

For constructing an observer in accordance with Theorem 5.3, the following conditions must be fulfilled:

(i) the pair $\{C, A\}$ is observable, the pair $\{A, B''\}$ is controllable,

(ii) $\operatorname{rank} C = l$, $\operatorname{rank} B'' = r < l$, $\operatorname{rank} CB'' = r$,

(iii) the invariant zeros defined by the Rosenbrock matrix

$$R(s) = \begin{pmatrix} sI - A & -B'' \\ C & 0 \end{pmatrix}$$

either are absent or lie in $\mathbb{C}_-$.

Taking into account the block structure of the matrices A and B'' in system (6.65) as well as the form $C = (I_{l\times l}; 0)$ for this system, we can write the indicated conditions in explicit form.

Let us consider in greater detail condition (i). By virtue of the Rosenbrock observability criterion (Theorem 2.11), the pair $\{C, A\}$ is observable if and only if the condition

$$\operatorname{rank} \begin{pmatrix} sI_n - A \\ C \end{pmatrix} = \operatorname{rank} \begin{pmatrix} sI_l - A_1 & -A_2 \\ -A_3 & sI_{n-l} - A_4 \\ I_l & 0 \end{pmatrix} = n$$

is fulfilled for all $s \in \mathbb{C}$. Obviously, this condition is fulfilled if and only if

$$\operatorname{rank} \begin{pmatrix} sI_{n-l} - A_4 \\ -A_2 \end{pmatrix} = n - l, \tag{6.67}$$

i.e., if the pair $\{A_2, A_4\}$ is observable. Note that $A_2 \in \mathbb{R}^{l\times(n-l)}$, and therefore, for condition (6.67) to be fulfilled, it is sufficient that the condition $\operatorname{rank} A_2 = n - l$ be fulfilled.

Let us now consider condition (ii). For the indicated structure of the matrix C and for $l > n - l$ the first two conditions are automatically fulfilled. The third condition, since $CB'' = B_2$, has the form

$$\operatorname{rank}(B_2) = \operatorname{rank} \begin{pmatrix} B_2 \\ B_4 \end{pmatrix}. \tag{6.68}$$

Note that if $\operatorname{rank} B_2 = n - l$, then, by virtue of the dimensions of the matrices $B_2 \in \mathbb{R}^{l\times(n-l)}$, $B'' \in \mathbb{R}^{n\times(n-l)}$ and the condition $l > n - l$, condition (6.68) is fulfilled for any matrix B_4.

Let us now consider condition (iii) for which purpose we write the Rosenbrock matrix in block form

$$R(s) = \begin{pmatrix} sI - A_1 & -A_2 & -B_2 \\ -A_3 & sI - A_4 & -B_4 \\ I & 0 & 0 \end{pmatrix}.$$

It is obvious that the invariant zeros of this matrix coincide with the invariant zeros of the matrix

$$R'(s) = \begin{pmatrix} sI - A_4 & -B_4 \\ -A_2 & -B_2 \end{pmatrix}.$$

For the original system to have no invariant zeros, it is required that the condition

$$\operatorname{rank} R'(s) = \operatorname{rank} \begin{pmatrix} sI - A_4 & -B_4 \\ -A_2 & -B_2 \end{pmatrix} = n - l + r \tag{6.69}$$

should be fulfilled for all $s \in \mathbb{C}$.

From the obtained conditions we have the following theorem.

Theorem 6.27. *Let* $\operatorname{rank} C = l$, $l > \frac{n}{2}$, *for the bilinear system* (6.45). *Then, by the nondegenerate change of coordinates, the matrices* A, B *and* C *can be reduced to block form* (6.66). *Also suppose that the following conditions are fulfilled*:

(i) *the pair* $\{A_2, A_4\}$ *is observable, and the pair* $\{A, B''\}$ *is controllable,*

(ii) $\operatorname{rank}(B_2) = \operatorname{rank} \begin{pmatrix} B_2 \\ B_4 \end{pmatrix} = r$ $(r \leq n - l)$,

(iii) *condition* (6.69) *is fulfilled for all* $s \in \mathbb{C}$ $(s \in \mathbb{C}_-)$.

Then, for system (6.45), *we can construct an exponential observer with any preassigned rate of convergence (with the rate of convergence defined by the invariant zeros of the Rosenbrock matrix from condition* (6.69)).

Conclusion

In Chap. 6 we considered the problem of synthesis of observers for one class of nonlinear systems, namely, for bilinear systems. We considered the conditions of observability of systems of this kind and obtained some sufficient conditions of uniform, with respect to $u(t)$, observability of these systems (Theorems 6.4, 6.14).

We considered in detail the case of planar bilinear systems where algorithms of synthesizing observers were proposed under different constraints imposed on the parameters of the system (Theorems 6.6–6.12)

For bilinear systems of arbitrary dimensions we considered different cases of degeneration of the bilinearity matrix (Theorems 6.15, 6.16) and also considered algorithms of synthesis of observers for systems with vector output (Theorems 6.19–6.27).

Chapter 7

Observers for discrete systems

The theory of observers for discrete systems is much similar to the theory of observers for continuous systems although it differs by certain specific features. Therefore, in this chapter we consider briefly only the main ideas and methods which were discussed in detain in the preceding chapters.

7.1 Mathematical models of discrete objects

In the theory of discrete observation of the state of stationary linear objects we deal with regressive models

$$y_{n+k} + a_n y_{n+k-1} + \cdots + a_1 y_k = b_{m+1} u_{m+k} + \cdots + b_1 u_k \tag{7.1}$$

or with dynamical models in the space of states

$$x_{k+1} = Ax_k + Bu_k, \quad y_k = Cx_k, \tag{7.2}$$

where u_k is the input of the object at the time moment $k = 0, 1, 2, \ldots$ and y_k is the output of the object, x_k is the phase vector or the vector of state of the object from $\mathbb{R}^n$; $(a_1, \ldots, a_n)$, $(b_1, \ldots, b_{m+1})$, or $\{A, B, C\}$ are parameters of the object, scalar or matrix respectively. Here n is the order of the object, $r = n - m$ is the relative order of the object, for physically realizable models satisfying cause and effect relations we always have $r \geq 1$.

Generally speaking, models (7.1) and (7.2) are not equivalent since (7.1) does not always imply (7.2).

Every model for the defined input sequence $\{u_k\}$ and the arbitrary initial state

$$(y_{n-1}, y_{n-2}, \ldots, y_0) \quad \text{or} \quad x_0 \in \mathbb{R}^n$$

generates a unique solution, namely, a discrete sequence of the output $\{y_k\}_0^\infty$ or the state $\{x_k\}_0^\infty$ respectively.

The discrete sequence $\{\xi_k\}_0^\infty$ which grows not faster than the degree of a certain positive number λ, i.e., $|\xi_k| \leq \lambda^k$, $k = 0, 1, 2, \ldots$, can be associated with the function of the complex variable $\xi(z)$, $z \in \mathbb{C}$, by means of the *Z-transformation* via the expression

$$\xi(z) = Z[\xi_k] = \sum_{k=0}^{\infty} \xi_k z^{-k} \quad \text{for} \quad |z| > \lambda.$$

After the substitution $z = e^{\Delta s}$, $\Delta = \text{const} > 0$, the Z-transformation passes into the so-called *Laplace discrete transformation.* It can be immediately seen from this formula that for obtaining the inverse Z^{-1}-transformation it suffices to decompose the function $\xi(z)$ according to the degrees z^{-k}. The coefficients in these degrees form the required sequence $\{\xi_k\}_0^{\infty}$.

By means of the Z-transformation the recurrence equation which connects the input and output of the object reduces to an algebraic equation. For equation (7.1) under zero initial conditions we have a relation

$$Y(z) = W(z)\,U(z), \tag{7.3}$$

where $Y(z) = Z[y_k]$, $U(z) = Z[u_k]$, and $W(z)$ is a transfer function of the object in the form of a ratio of two polynomials[1]

$$W(z) = \frac{\alpha_n(z)}{\beta_m(z)} = \frac{(b_{m+1}z^m + \cdots + b_1)}{(z^n + a_n z^{n-1} + \cdots + a_1)}. \tag{7.4}$$

By analogy, the transfer function is defined for an object given in the space of states. From (7.2) under the zero initial condition $x_0 = 0$ we have

$$W(z) = C(zE - A)^{-1}B. \tag{7.5}$$

For the so-called scalar object the output y and input u of the object are scalars and its transfer function is a scalar function of the complex variable z. Otherwise, $W(z)$ is a matrix transfer function.

For a scalar object formula (7.3) makes it possible to find a simple transition from (7.1) to (7.2). Indeed, from the relation

$$W(z) = \frac{\beta(z)}{\alpha(z)}$$

and equation (7.3) we can obtain a relation

$$\frac{Y(z)}{\beta(z)} = \frac{U(z)}{\alpha(z)}.$$

Employing this relation, we can introduce a scalar variable x_k^1 using the formula

$$X^1(z) = \frac{Y(z)}{\beta(z)} = \frac{U(z)}{\alpha(z)},$$

[1]Below, where it is not essential, we omit the subscript of the polynomial which indicates its degree.

where $X^1(z) = Z\left[x_k^1\right]$. The latter is equivalent to the two equations

$$\begin{cases} \alpha(z)X^1(z) = U(z) \\ \qquad Y(z) = \beta(z)X^1(z). \end{cases} \tag{7.6}$$

Using now the inverse transformation Z^{-1}, we can pass from (7.6) first to the regressive equations

$$\begin{aligned} x_{n+k}^1 + a_n x_{n+k}^1 + \cdots + a_1 x_k^1 &= u_k \\ y_k = b_{m+1} x_{m+k}^1 + \cdots + b_1 x_k^1 \end{aligned} \tag{7.7}$$

and then, with the use of the new state variables $x^1, x^2, x^3, \ldots, x^n$ related as

$$\begin{cases} x_{k+1}^1 = x_k^2 \\ x_{k+1}^2 = x_k^3 \\ \quad\vdots \\ x_{k+1}^{n-1} = x_k^n, \end{cases} \tag{7.8}$$

to equations in the space of states

$$\begin{aligned} x_{k+1}^n &= -\sum_{i=1}^n a_i x_k^i + u_k \\ y_k &= \sum_{i=1}^n b_i x_k^i, \quad b_i = 0 \quad \text{for} \quad i > m+1. \end{aligned} \tag{7.9}$$

The use of the vector-matrix notation

$$A = \begin{pmatrix} 0 & 1 & 0 & \ldots & 0 \\ 0 & 0 & 1 & \ldots & 0 \\ & \vdots & & \ldots & \vdots \\ 0 & 0 & 0 & \cdots & 1 \\ -a_1 & -a_2 & -a_3 & \cdots & -a_n \end{pmatrix}, \quad b = \begin{pmatrix} 0 \\ 0 \\ \vdots \\ 0 \\ 1 \end{pmatrix},$$
$$c = (b_1, b_2, \ldots, b_{m+1}, 0, \ldots, 0)$$

makes the form of equations (7.8), (7.9) identical to that of equations (7.2). For this reason, it makes sense to consider in the sequel models defined in the space of state.

7.2 Discrete observability and observers. Canonical forms

We shall begin with the following definition.

Definition 7.1. The dynamical discrete system (the recurrence equation)

$$(NS)\quad \begin{cases} x_{k+1} = f(x_k), & x \in \mathbb{R}^n \\ y_k = h(x_k), & y \in \mathbb{R}^l, \quad k = 0, 1, 2, \dots \end{cases}$$

is said to be

- *observable* if the finite output sequence $\{y_0, y_1, \dots, y_{N-1}\}$, $N \geq n$, makes it possible to reconstruct the initial state x_0 of the system,
- *uniformly observable* with respect to $k \geq 0$ if the reconstruction is possible with the use of the finite sequence $\{y_k, y_{k+1}, \dots, y_{k+(N-1)}\}$ for any k.

Let us consider a scalar linear stationary S_0-system with zero input

$$(S_0)\quad \begin{cases} x_{k+1} = Ax_k \\ y_k = cx_k, \quad k = 0, 1, 2, \dots \end{cases}$$

and write the obvious relations

$$\begin{cases} y_0 = cx_0 \\ y_1 = cAx_0 \\ \quad\vdots \\ y_{N-1} = cA^{N-1}x_0 \end{cases}$$

which form a system of linear algebraic equations for the unknown vector x_0 of the initial conditions. In accordance with the Cayley–Hamilton theorem we can restrict ourselves to the first n relations, i.e., $N = n$. Then the solvability of the obtained system of equations is connected with the invertibility of the matrix

$$N(c, A) = \begin{pmatrix} c \\ cA \\ \vdots \\ cA^{n-1} \end{pmatrix} \tag{7.10}$$

which is known as the *observability matrix*. Thus, the S_0-system is observable if and only if

$$\operatorname{rank} N(c, A) = n. \tag{7.11}$$

Relation (7.11) defines the criterion of observability in the form of the *Kalman–Krasovskii rank condition*. This result is also valid for the standard S-system

$$(S)\quad \begin{cases} x_{k+1} = Ax_k + bu_k, \\ y_k = cu_k, \quad k = 0, 1, 2, \dots, \end{cases}$$

for the known input u_k.

For S-systems with vector observers, i.e., when $y \in \mathbb{R}^l$ and $l > 1$,

$$\begin{cases} x_{k+1} = Ax_k + bu_k \\ y_k = Cu_k \end{cases}$$

with an additional natural assumption that

$$\operatorname{rank} C = l,$$

the criterion of observability assumes the form

$$\operatorname{rank} N(C, A) = \operatorname{rank} \begin{pmatrix} C \\ CA \\ \vdots \\ CA^{n-l} \end{pmatrix} = n. \tag{7.12}$$

The nondegenerate change of coordinates

$$\xi = Mx, \quad \det M \neq 0,$$

preserves the observability since

$$N(CM^{-1}, MAM^{-1}) = N(C, A)(M^{-1}),$$

where $N(CM^{-1}, MAM^{-1})$ is an observability matrix for the transformed system

$$\begin{cases} \xi_{k+1} = MAM^{-1}\xi_k \\ y_k = CM^{-1}\xi_k. \end{cases}$$

As in a continuous case, the *observability index* ν for the observable pair $\{C, A\}$ is defined as a minimal number for which the rank condition

$$\operatorname{rank} N_\nu(C, A) = \operatorname{rank} \begin{pmatrix} C \\ CA \\ \vdots \\ CA^{\nu-1} \end{pmatrix} = n$$

is fulfilled. For ν the estimate $\nu \leq n - l + 1$ holds, for $l = 1$ we have a relation $\nu = n$.

For discrete linear stationary observable systems we have the same canonical forms of observability as for continuous systems (described in detail in Chap. 2). In particular, the scalar S-systems can be reduced, by means of a nondegenerate change of coordinates, to the *first canonical form of observability* $\{c, A, b\}$:

$$c = (1, 0, \ldots, 0),$$

$$A = \begin{pmatrix} 0 & 1 & 0 & \dots & 0 \\ 0 & 0 & 1 & \dots & 0 \\ \dots & \dots & \dots & \dots & 1 \\ -a_1 & -a_2 & -a_3 & \dots & -a_n \end{pmatrix}, \quad b = \begin{pmatrix} cb \\ \vdots \\ cA^{n-1}b \end{pmatrix},$$

or to the *second canonical form of observability* $\{c, A, b\}$:

$$c = (0, \dots, 0, 1), \quad A = \begin{pmatrix} 0 & \dots & 0 & -a_1 \\ 1 & \dots & 0 & -a_2 \\ \vdots & \ddots & \vdots & \vdots \\ 0 & \dots & 1 & -a_n \end{pmatrix}, \quad b = \begin{pmatrix} b_1 \\ b_2 \\ \vdots \\ b_n \end{pmatrix}.$$

The observability of the canonical pairs $\{c, A\}$ can be easily verified.

The first and second canonical forms of observability and the Luenberger canonical form also hold for vector S-systems (see Chap. 2). The methods of reducing systems to canonical forms completely coincide with the corresponding methods for continuous systems. We omit the details.

By analogy with a continuous case, we introduce the notion of detectability.

Definition 7.2. The dynamical discrete system

$$(NS) \quad \begin{cases} x_{k+1} = f(x_k), & x \in \mathbb{R}^n \\ y_k = h(x_k), & y \in \mathbb{R}^l, \ k = 0, 1, 2, \dots, \end{cases}$$

is said to be *detectable* if the observable (latent) dynamics is asymptotically stable.

Let us consider in greater detail the linear stationary S-system

$$(S) \quad \begin{cases} x_{k+1} = Ax_k + Bu_k \\ y_k = Cx_k, \ k = 0, 1, 2, \dots . \end{cases}$$

Suppose that the pair $\{C, A\}$ is observable (not fully observable). Then we have a condition

$$\operatorname{rank} N(C, A) = q < n.$$

In this case, by means of the similarity transformation M ($\det M \neq 0$)

$$\xi = Mx,$$

we can reduce the S-system to a set of subsystems S^1 and S^2, where

$$(S^1): \quad \xi^1_{k+1} = A_{11}\xi^1_k + B^1 u_k$$
$$(S^2): \quad \xi^2_{k+1} = A_{21}\xi^1_k + A_{22}\xi^2_k + B^2 u_k$$
$$y_k = C^1 \xi^1_k.$$

Here the vector $\xi^1 \in \mathbb{R}^q$, the vector $\xi^2 \in \mathbb{R}^{n-q}$, the matrices A_{11}, A_{21}, A_{22},B^1, B^2, and C^1 have the requisite dimensions, and the pair $\{C^1, A_{11}\}$ is observable.

Thus, the nonobservable variables ξ^2 are isolated, as a result of the transformation, into a S^2-subsystem of "latent" motions. The variables ξ^2 may depend on the observable variables ξ^1 and on the input u.

The transfer function of the system can be written with the use of the initial equation (S)

$$W(z) = c(zE - A)^{-1}b = \frac{\beta(z)}{\alpha(z)}$$

or with the use of the equation for subsystems (S^1), (S^2)

$$W_1(z) = c^1(zE - A_{11})^{-1}b^{-1} = \frac{\beta^1(z)}{\alpha^1(z)}.$$

The expression $W_1(s)$ follows from $W(s)$ after cancelling the $(n-q)$ common zeros of the polynomials $\beta(z)$ and $\alpha(z)$. This degeneration allows us to lower the order of the S-system to q. Physically this cancelling of zeros and poles of the transfer function is justified if they are stable, i.e., lie in a unit circle. The detectability implies that A_{22} is a Hurwitz matrix.

Along with the problem of observation for discrete systems we shall consider a stabilization problem. Stabilizability is one of the main properties of a controlled object. We can give the following definition.

Definition 7.3. The system (*NS*)

$$(NS) \quad \begin{cases} x_{k+1} = f(x_k, u_k), \quad f(0,0) = 0 \\ y_k = h(x_k),\ k = 0, 1, 2, \dots, \end{cases}$$

where $h(\cdot)$ and $f(\cdot,\cdot)$ are some functions defined in $\mathbb{R}^n$, is said to be *stabilizable*[2] *at zero of* $\mathbb{R}^n$ if there exists a feedback

$$u_k = u(y_k, x_k)$$

such that the closed control system

$$\begin{cases} x_{k+1} = f(x_k, u(y_k, x_k)) \\ y_k = h(x_k) \end{cases}$$

is asymptotically stable at zero.

---

[2]Here and in what follows a system is stabilizable at zero, and we can do without stipulating this.

Here and in what follows we understand the symbol $u(y,x)$ as a feedback dependent on each variable or the values of the variables at the preceding time moments. This dependence may be static or dynamical respectively, the constructed feedback is static or dynamical.

In particular, the S-system

$$\begin{cases} x_{k+1} = Ax_k + bu_k \\ y_k = cx_k \end{cases}$$

is stabilizable at zero by its state if for the feedback

$$u_k = -qx_k, \quad q = (q_1, \dots, q_n),$$

the closed system

$$x_{k+1} = (A - bq)x_k$$

is asymptotically stable. To put it otherwise, $A_q = A - bq$ is a Hurwitz matrix.

The sufficient condition of stabilizability is the condition of controllability of the system of object (NS) which is introduced by the following definition.

Definition 7.4. The system

$$(NS) \qquad \begin{cases} x_{k+1} = f(x_k, u_k) \\ y_k = h(x_k) \end{cases}$$

is *controllable with respect to its state* in $\mathbb{R}^n$ if, for any pair of points x^1, x^2 in $\mathbb{R}^n$, there exists a sequence of inputs u_0, u_1, ..., u_{N-1} which reduces the system (*NS*) from the state x^1 to the state x^2 in a finite time interval N.

In order to analyze the controllability of the state of the S-system, we write the following chain of obvious relations

$$\begin{aligned} x_1 &= Ax_0 + bu_0 \\ x_2 &= A^2x_0 + Abu_0 + bu_1 \\ &\vdots \\ x_N &= A^N x_0 + A^{N-1}bu_0 + \dots + bu_{N-1}. \end{aligned}$$

Setting $x_0 = x^1$, $x_N = x^2$, we obtain from the last relations an equation

$$x^2 = A^N x^1 + \left(A^{N-1}b, A^{N-2}, \dots, b\right) \begin{pmatrix} u_0 \\ u_1 \\ \vdots \\ u_{N-1} \end{pmatrix}$$

which must be solvable for the unknowns $u_0, u_1, \ldots, u_N$ for any x^1 and x^2 for which purpose the vectors $A^{N-1}b,\ A^{N-2}b,\ \ldots,\ b$ must form a basis in $\mathbb{R}^n$. Thus, the S-system is controllable if and only if

$$\operatorname{rank} K(A,b) = \operatorname{rank}(b, Ab, \ldots, A^{n-1}b) = n. \tag{7.13}$$

This condition of controllability is known as a *Kalman–Krasovskii rank criterion*. If it is fulfilled, the pair $\{A, b\}$ is said to be controllable.

If a linear system is a vector system, i.e., $u \in \mathbb{R}^m$, $y \in \mathbb{R}^l$,

$$\begin{cases} x_{k+1} = Ax_k + Bu_k \\ \quad y_k = Cx_k, \end{cases}$$

where the matrix B is of the maximal rank m, then the Kalman–Krasovskii criterion differs from (7.13) by the fact that the matrix $K(A, B)$ is a rectangular $n \times [m(n-m+1)]$-matrix, but, as before,

$$\operatorname{rank} K(A, B) = \operatorname{rank}(B, AB, \ldots, A^{n-m}B) = n.$$

For the controllable pair $\{A, B\}$ the controllability index μ is defined as the minimal number for which

$$\operatorname{rank} K_\mu(A, B) = \operatorname{rank}(B, AB, \ldots, A^{\mu-1}B) = n.$$

In this case, $\mu \le n - m + 1$ and if $m = 1$, then $\mu = n$.

If the pair $\{A, B\}$ is noncontrollable (not fully controllable),

$$\operatorname{rank} K(A, B) = p < n,$$

then there exists a similarity transformation

$$\xi = \begin{pmatrix} \xi^1 \\ \xi^2 \end{pmatrix} = Mx$$

which splits the S-system into two subsystems S_1 and S_2 with the motion equations

$$\begin{aligned} (S_1) \quad & \xi^1_{k+1} = A_{11}\xi^1_k \\ (S_2) \quad & \xi^2_{k+2} = A_{21}\xi^1_k + A_{22}\xi^2_k + B^2 u_k \\ & y = (C^1, C^2)\begin{pmatrix} \xi^1 \\ \xi^2 \end{pmatrix}, \end{aligned}$$

where $(n - p)$ is the dimension of the system S_1 and p is the dimension of the system S_2, A_{11}, A_{21}, A_{22}, B^2, C^1, C^2 are matrices of the corresponding dimensions, with the pair $\{A_{22}, B^2\}$ being controllable.

We can see from these equations that the control u does not affect the component ξ^1 of the vector ξ, and, consequently, the S-system is stabilizable if and only if A_{11} is a Hurwitz matrix. To put it otherwise, the part of the dynamical system which cannot be controlled must be asymptotically stable.

Thus, in an arbitrary S-system we can isolate the following subsystems:
(i) a controllable and observable subsystem (S^1),
(ii) a controllable but not observable subsystem (S^2),
(iii) a noncontrollable and nonobservable subsystem (S^3),
(iv) a nonobservable and noncontrollable subsystem (S^4).

This decomposition of a S-system is known as a *Kalman decomposition*; the S^1-system corresponds to the so-called *minimal realization of the S-system* which has physical meaning when the eigenmotions of the S^3- and S^4-subsystems are asymptotically stable. The following motion equations correspond to the Kalman decomposition of the S-system:

$$(S^1)\quad \xi^1 = A_{11}\xi_k^1 + A_{13}\xi_k^3 + B^1 u_k$$

$$(S^2)\quad \xi^2 = A_{21}\xi_k^1 + A_{22}\xi_k^2 + A_{23}\xi_k^3 + A_{24}\xi_k^4 + B^2 u_k$$

$$(S^3)\quad \xi_{k+1}^3 = A_{33}\xi_k^3$$

$$(S^4)\quad \xi_{k+1}^4 = A_{43}\xi_k^3 + A_{44}\xi_k^4$$

$$y = (C^1, 0, C^3, 0)\begin{pmatrix}\xi_1\\ \vdots\\ \xi_4\end{pmatrix} = C^1\xi^1 + C^3\xi^3,$$

where $\xi^i \in \mathbb{R}^{n_i}$ with $n_1 \leq \min(p,q)$; $n_1+n_2 = p$; $n_1+n_3 = q$; $n_1+n_2+n_3+n_4 = n$.

It is obvious that the case of the general position is associated only with the minimal realization of the S-system.

Controllability and observability, stabilizability and detectability are *dually* connected, namely, the expressions for the matrices of controllability and observability of the S-system immediately imply identities

$$\begin{cases} K(A,B) \equiv N^\top(B^\top, A^\top) \\ N(C,A) \equiv K^\top(A^\top, C^\top). \end{cases}$$

The left-side matrices are connected with the system

$$(S)\qquad \begin{cases} x_{k+1} = Ax_k + Bu_k \\ y_k = Cx_k \end{cases}$$

and the right-side matrices with the $S^\top$-system

$$(S^\top) \quad \begin{cases} \xi_{k+1} = A^\top \xi_k + C^\top \xi_k \\ \xi_k = B^\top \xi_k \end{cases}$$

known as a *dual system*. The indicated identities introduce the so-called duality relations which give, in particular, two significant inferences.

1°. The S-system is controllable (observable) if and only if the $S^\top$-system is observable (controllable).

2°. The S-system is stabilizable (detectable) if and only if the $S^\top$-system is detectable (stabilizable).

The notions of observability and detectability are convenient for estimating the state of a dynamical system from the measurements of the output of the system.

The following definition introduces the notion of a discrete observer.

Definition 7.5. The dynamical system

$$\hat{x}_{k+1} = F(\hat{x}_k, y_k)$$

is said to be an *asymptotic observer* of an NS-system of the form

$$\begin{cases} x_{k+1} = f(x_k) \\ y_k = h(x_k), \quad k = 0, 1, 2, \dots, \end{cases}$$

if

$$\lim_{k\to\infty} \|x_k - \hat{x}_k\| = 0.$$

If there exists a number k^* such that

$$\hat{x}_k = x_k$$

for all $k \geq k^*$, then the observer is said to be finite.

For S-systems the problem of estimating the phase vector is solved by the linear observer

$$\hat{x}_{k+1} = A\hat{x}_k - L(C\hat{x}_k - y_k) = A_L\hat{x}_k + LCy_k, \tag{7.14}$$

where L is the $(n \times l)$ feedback matrix of the observer. The estimation error $\varepsilon = x - \hat{x}$ satisfies the equation

$$\varepsilon_{k+1} = A_L\varepsilon_k$$

and the observer solves the estimation problem asymptotically if A_L is a Hurwitz matrix or finitely if A_L is a nilpotent matrix. As was shown in Chap. 2, if the pair $\{C, A\}$

is observable, the spectrum of the observer spec$\{A_L\}$ may be defined arbitrarily. If the pair $\{C, A\}$ is detectable, then the part spec$\{A_L\}$ of the spectrum of the observer is fixed and coincides with the spectrum of the nonobservable subsystem and the remaining part of the spectrum spec$\{A_L\}$ is defined arbitrarily (it stands to reason that the arrangement of the spectrum relative to the real axis must be symmetrical).

The dimension of the observer can be diminished by the dimension of the output if rank $C = l$. Observers of lowered order of this kind are traditionally called *Luenberger observers* on the basis of the following statement.

Lemma 7.6. *Suppose that in the vector discrete S_0-system*

$$\begin{cases} x_{k+1} = Ax_k \\ y_k = Cx_k, \quad k = 0, 1, \dots, \quad x \in \mathbb{R}^n, \quad y \in \mathbb{R}^l, \end{cases}$$

the pair $\{C, A\}$ is observable. Then, by a nonsingular transformation this system can be reduced to the form

$$\begin{cases} x'_{k+1} = A_{11}x'_k + A_{12}y_k \\ y_{k+1} = A_{21}x'_k + A_{22}y_k, \quad k = 0, 1, \dots, \quad x' \in \mathbb{R}^{n-l}, \quad y \in \mathbb{R}^l, \end{cases} \tag{7.15}$$

where the $(n-l) \times (n-l)$-matrix A_{11} has any predefined spectrum.

A constructive proof of this lemma can be obtained with the use of the method of pseudoinputs (see Sec. 5.3), we do not give it here. This lemma can also be generalized to the case of detectability of the pair $\{C, A\}$.

Under the conditions of Lemma 7.6 the observers which solve the observation problem for S_0-systems are given by the equations

$$\bar{x}'_{k+1} = A_{11}\bar{x}'_k + A_{12}y_k \tag{7.16}$$

$$\bar{x}_k = M \begin{pmatrix} \bar{x}'_k \\ y_k \end{pmatrix}, \tag{7.16$'$}$$

where $\bar{x}' \in \mathbb{R}^{n-l}$ and M is the transformation matrix which is mentioned in Lemma 7.6. The estimation error $\varepsilon = \bar{x} - x'$ is defined by the equation

$$\varepsilon_{k+1} = A_{11}\varepsilon_k, \quad k = 0, 1, 2, \dots, \tag{7.17}$$

and, for the requisite choice of the transformation matrix which is mentioned in the conditions of the lemma the error ε_k asymptotically or finitely becomes zero.

Note that equations (7.16)–(7.16$'$) for the observer are also valid for the case of the detectability of the pair $\{C, A\}$, the only difference is that the spectrum of the matrix A_{11} has a fixed Hurwitz component of the latent dynamics.

We have thus shown that the dimension of the observer which reconstructs the full-phase vector cannot be smaller than the number $r = n - l$, where l is the dimension of the output.

Note that for a vector system an observer of any intermediate dimension higher than $(n - l)$ can be constructed according to the same scheme. This fact is based on the following lemma.

Lemma 7.7. *Suppose that in the vector discrete S_0-system*

$$\begin{cases} x_{k+1} = Ax_k, \quad k = 0, 1, \dots \\ y_k = Cx_k, \quad x \in \mathbb{R}^n, \quad y \in \mathbb{R}^l, \end{cases}$$

rank $C = l$, *and the pair $\{C, A\}$ is observable. Then, for any number ρ, $1 < \rho < l$, there exists a ρ-dimensional output $y^\rho = C_\rho y \in \mathbb{R}^\rho$ and a similarity transformation $\binom{x'}{y^\rho} = M_\rho x$ such that $(n - \rho) \times (n - \rho)$ is the matrix A_{11} in the transformed system*

$$\begin{cases} x'_{k+1} = A_{11}x'_k + A_{12}y^\rho_k \\ y^\rho_{k+1} = A_{21}x'_k + A_{22}y^\rho_k, \quad k = 0, 1, \dots . \end{cases} \tag{7.18}$$

This matrix is a Hurwitz matrix and has a predefined spectrum (here $x' \in \mathbb{R}^{n-\rho}$).

In this case, the observation problem is solved by the observer

$$\begin{cases} \bar{x}'_{k+1} = \bar{A}_{11}\bar{x}'_k + A_{12}y^\rho_k \\ \bar{x}_k = M_\rho^{-1} \begin{pmatrix} \bar{x}'_k \\ y^\rho_k \end{pmatrix} \end{cases}$$

of dimension $(n - \rho)$.

It is useful to note that under the condition rank $C = l$ for system (7.18) the $(l - \rho)$-dimensional output $y^{l-\rho} = C_{l-\rho}x'$ is also defined (it is defined with the use of the output $y \in \mathbb{R}^l$ and $y^\rho \in \mathbb{R}^\rho$). Then we can rewrite equation (7.18) as a state equation with the known external signal y^ρ and the new output $y^{1-\rho}$, i.e., as a set of equations

$$\begin{cases} x'_{k+1} = A_{11}x'_k + A_{12}y^\rho_k, \quad k = 0, 1, \dots \\ y^{l-\rho}_k = C_{n-\rho}x'_k. \end{cases} \tag{7.19}$$

It is important to emphasize here that under the conditions of Lemma 7.7 the pair of matrices $\{C_{l-\rho}, A_{11}\}$ is observable. This fact will be used in the synthesis of observers under the conditions of uncertainty.

Note that if we deal with a controllable S-system, i.e., with a system which has an input, then, for this system, instead of (7.14), we take an observer in the form

$$\hat{x}_{k+1} = A\hat{x}_k - L(C\hat{x}_k - y_k) + Bu_k. \tag{7.20}$$

7.3 Method of pseudoinputs in the problem of synthesis of functional observers

Consider a problem of estimation of a p-dimensional linear functional

$$\sigma = Fx \tag{7.21}$$

from the measurements of the l-dimensional output ($l < n$)

$$y_k = Cx_k \tag{7.22}$$

of the stationary discrete system

$$x_{k+1} = Ax_k, \quad k = 0, 1, 2, \ldots, \tag{7.23}$$

using an observer of the order as minimal as possible.

Here all matrices are assumed to be known, F and C are matrices of full rank, and the pairs $\{C, A\}$ and $\{F, A\}$ are observable.

7.3.1 Scalar system, scalar functional

As we did in Chap. 4, we shall begin with considering the case of a scalar functional and a scalar observer, i.e., where $l = p = 1$, and use the method of pseudoinputs for solving the problem. This means that along with (7.23) we also consider a system

$$x_{k+1} = Ax_k + Lv_k; \quad k = 0, 1, 2, \ldots, \tag{7.24}$$

with a "pseudoinput" v_k and with a vector of the input L. The latter approach is only a convenient methodical technique and makes it possible to introduce, in a natural way, two transfer functions

$$\begin{aligned} W_y(z) &= C(zI - A)^{-1}L = \frac{\beta_y(z)}{\alpha(z)}, \\ W_\sigma(z) &= F(zI - A)^{-1}L = \frac{\beta_\sigma(z)}{\alpha(z)} \end{aligned} \tag{7.25}$$

with the same polynomials in the denominator which coincide with the characteristic polynomial of the matrix A, i.e.,

$$\alpha(z) = \det(zI - A) = z^n + a_n z^{n-1} + \cdots + a_1,$$

and, in general, different polynomials in the numerator in (7.25) (we can naturally assume that $C \nparallel F$).

By virtue of the assumptions that we have made, the polynomials $\beta_y(z)$, $\beta_\sigma(z)$ and $\alpha(z)$, are coprime, we can also achieve the same for the polynomials $\beta_y(z)$ and $\beta_\sigma(z)$ by a requisite choice of the vector L.

Under these conditions, the transfer function $W_0(z)$ of the required observer

$$\sigma = W_0(z)y \tag{7.26}$$

is embedded into the family of transfer functions defined by the ratio of polynomials $\beta_\sigma(z)$ and $\beta_y(z)$, i.e.,

$$W_0(z) = \frac{\beta_\sigma(z)}{\beta_y(z)}. \tag{7.27}$$

The physically realizable transfer functions of functional observers form in this family of transfer functions a subset distinguished by the conditions

$$\begin{cases} \deg \beta_\sigma(z) \le \deg \beta_y(z), \\ \beta_y(z) \quad \text{is a Hurwitz polynomial.} \end{cases} \tag{7.28}$$

Since the order of the required observer is defined by the number $\deg \beta_y(z)$, we have to isolate, in the subset (7.28), transfer functions (7.27) for which the degree $\deg \beta_y(z)$ is minimal.

It should be emphasized that the problem of minimization of the degree $\deg \beta_y(z)$ of the polynomial is solved under the degree restriction (7.28) and is, therefore, a non-classical problem of optimization.

Let us formulate this problem in a more customary algebraic form, for which purpose we note that the polynomials $\beta_\sigma(z)$ and $\beta_y(z)$ are of a degree not higher than $(n-1)$ and can be written with the use of Markov parameters of the triplets $\{F, A, L\}$ and $\{C, A, L\}$ respectively by the expressions

$$\begin{aligned} \beta_\sigma(z) &= z^{n-1}(FL) + z^{n-2}(FAL + a_n FL) \\ &\quad + z^{n-3}(FA^2L + a_n FAL + a_{n-1} FL) + \cdots \\ \beta_y(z) &= z^{n-1}(CL) + z^{n-2}(CAL + a_n CL) \\ &\quad + z^{n-3}(CA^2L + a_n CAL + a_{n-1} CL) + \cdots \end{aligned} \tag{7.29}$$

where $a_1, a_2, \ldots, a_n$ are parameters of the characteristic polynomial

$$\alpha(z) = z^n + a_n z^{n-1} + \cdots + a_1.$$

Suppose that $\varkappa$ is the required minimal degree of the Hurwitz polynomial $\beta_y(z)$ which provides a solution of the problem under consideration, and then, obviously, the relations

$$\begin{aligned} CL &= CAL = \cdots = CA^{n-\varkappa-2}L = 0 \\ FL &= FAL = \cdots = FA^{n-\varkappa-2}L = 0 \end{aligned} \tag{7.30}$$

should hold simultaneously, with

$$CA^{n-\varkappa-1}L \ne 0. \tag{7.31}$$

Thus, the polynomial $\beta_y(z)$ responsible for stability of the observer is defined by the equation

$$\beta_y^{\varkappa}(z) = z^{\varkappa}(CA^{n-\varkappa-1}L) + z^{\varkappa-1}(CA^{n-\varkappa}L + a_n CA^{n-\varkappa-1}L) + \cdots \tag{7.32}$$

and, without loss of generality, we can set

$$CA^{n-\varkappa-1}L = 1.$$

Then, instead of saying that $\beta_y^{\varkappa}(z)$ is a Hurwitz polynomial, we can say that the column

$$L^{\varkappa} = \begin{pmatrix} L' \\ 1 \end{pmatrix} = (\dots, CA^{n-\varkappa}L + a_n, 1)^{\top} \tag{7.33}$$

is a Hurwitz column consisting of coefficients of the polynomial $\beta_y^{\varkappa}(z)$ (here $L^{\varkappa} \in \mathbb{R}^{\varkappa+1}$, $L' \in \mathbb{R}^{\varkappa}$).

The determining of the conditions of solvability of this problem and the synthesis of the observer can be conveniently carried out in a special basis with a minimal number of free parameters. Such a natural basis is the basis in which the pair $\{C, A\}$ has a canonical form of observability, i.e.,

$$c = (0, \dots, 0, 1), \quad A = \begin{pmatrix} 0 & \dots & 0 & -a_1 \\ 1 & \dots & 0 & -a_2 \\ \vdots & \ddots & \vdots & \vdots \\ 0 & \dots & 1 & -a_n \end{pmatrix},$$

and we shall suppose that in this basis the vectors F and L have components

$$F = (f_1, f_2, \dots, f_n), \quad L = (l_1, l_2, \dots, l_n)^{\top}.$$

Then it follows from the first relations in (7.30) that

$$l_n = l_{n-1} = \dots = l_{\varkappa+2} = 0,$$

i.e., the vector L has the structure

$$L = \begin{pmatrix} L^{\varkappa} \\ 0 \end{pmatrix} = (l_1, l_2, \dots, l_{\varkappa}, 1, 0, \dots, 0)^{\top},$$

and the set $(n - \varkappa - 1)$ of the second relations in (7.30) is equivalent to the system of linear equations

$$H_{\varkappa} l^{\varkappa} = \begin{pmatrix} f_1 & f_2 & \dots & f_{\varkappa} \\ f_2 & f_3 & \dots & f_{\varkappa+1} \\ \dots & \dots & \dots & \dots \\ f_{n-\varkappa-1} & f_{n-\varkappa} & \dots & f_{n-2} \end{pmatrix} \begin{pmatrix} l_1 \\ l_2 \\ \vdots \\ l_{\varkappa} \end{pmatrix} = - \begin{pmatrix} f_{\varkappa+1} \\ f_{\varkappa+2} \\ \vdots \\ f_{n-1} \end{pmatrix} = -h_{\varkappa}, \tag{7.34}$$

which, of course, is equal to system (4.18) in Chap. 4.

Since the components $(l_1, l_2, \dots, l_\varkappa)^\top$ of the vector are the parameters of the polynomial $\beta_y^\varkappa(z)$, i.e.,

$$\beta_y^\varkappa(z) = z^\varkappa + l_\varkappa z^{\varkappa-1} + \cdots + l_1,$$

the presence, for a certain $\varkappa$, of a Hurwitz solution of system (7.34) is a necessary and sufficient condition of existence of a functional observer of order $\varkappa$. In the terms of Z-transformation the equation for such an observer which forms an estimate $\bar{\sigma}$ of the functional σ has the form

$$\bar{\sigma} = \frac{\beta_\sigma(z)}{\beta_y^\varkappa(z)} y, \tag{7.35}$$

and the estimation errors

$$\varepsilon = \bar{\sigma} - \sigma$$

satisfy respectively the equation

$$\beta_y^\varkappa(z)\varepsilon = 0,$$

and this solves the observation problem since $\beta_y^\varkappa(z)$ is a Hurwitz polynomial.

The question concerning the minimal order $\varkappa^*$ of the functional observer (7.35) remains open. The computation of $\varkappa^*$ is possible in the framework of the following iteration procedure:

- first, from the equation

$$\operatorname{rank} H_\varkappa = \operatorname{rank}(H_\varkappa \;\; h_\varkappa) \tag{7.36}$$

  we find the minimal number $\varkappa_m$ for which system (7.34) is solvable; if, for this $\varkappa_m$, among the solutions of (7.34) there exists a Hurwitz vector $L^{\varkappa_m}$,

$$\varkappa^* = \varkappa_m,$$

  then the problem is solved,

- otherwise, we must increase $\varkappa_m$ by unity, and, since for any $\varkappa > \varkappa_m$ condition (7.36) is fulfilled, we should repeat the procedure; after a finite number of iterations we will find the required minimal order $\varkappa^*$ of the functional observer,

- the formula

$$\bar{\sigma} = \frac{\beta_\sigma(z)}{\beta_y^{\varkappa^*}(z)} y$$

  should be used to synthesize the required observer.

We have thus proved the following theorem.

Theorem 7.8. *For the observable n-dimensional system with a scalar output*

$$x_{k+1} = Ax_k, \quad y_k = Cx_k, \quad k = 0, 1, 2, \ldots,$$

and a linear scalar functional

$$\sigma = Fx$$

such that $F \nparallel C$ and the pair $\{F, A\}$ is observable, there exists a functional observer of order $\varkappa$ of form (7.35) *which reconstructs the functional σ if and only if the system of linear equations defined in the canonical basis of observability by relation* (7.34)

$$H_\varkappa l^\varkappa = -h_\varkappa,$$

has a solution $l^\varkappa$ such that $L^\varkappa = \binom{l^\varkappa}{1}$ is a Hurwitz vector.

Remark 7.9. The requirement that the pairs $\{C, A\}$ and $\{F, A\}$ should be in the general position is not necessary, the detectability of these pairs is sufficient for solving the problem.

7.3.2 Scalar observations, vector functional

The synthesis of a functional observer of the smallest order by the method of pseudoinputs for systems (7.22) and (7.23) can be generalized, without any essential changes, to p-dimensional vector functionals

$$\sigma = Fx$$

where, it stands to reason, it is relevant to suppose that

$$\operatorname{rank} F = p, \quad \operatorname{rank}\begin{pmatrix} F \\ C \end{pmatrix} = p + 1, \quad p > 1. \tag{7.37}$$

For the measurable output y_k of the system and each i-component $\sigma^i = F^i x$ ($i = 1, 2, \ldots, p$) of the functional σ we shall determine the transfer functions from the pseudoinputs v (see (7.24)) to the corresponding output y or σ_i, i.e.,

$$\sigma^i = F_i(zI - A)^{-1}Lv = \frac{\beta_\sigma^i(z)}{\alpha(z)}v, \quad i = 1, 2, \ldots, p, \tag{7.38}$$

$$y = C(zI - A)^{-1}Lv = \frac{\beta_y(z)}{\alpha(z)}v. \tag{7.39}$$

This makes it possible to determine, in the general form, the transfer vector function of the observer, i.e., transfer functions from y to each component of the functional, in the form

$$\sigma^i = \frac{\beta_\sigma^i(z)}{\beta_y(z)}y, \quad i = 1, 2, \ldots, p. \tag{7.40}$$

If we use now the vector of the pseudoinput L so that the conditions

$$\deg \beta_\sigma^i \le \deg \beta_y \quad \text{for all} \quad i = 1, 2, \dots, p, \tag{7.41}$$
$$\beta_y(z) \ \text{ is a Hurwitz polynomial,}$$

will be fulfilled, then the observation problem is solved by the observer with a vector transfer function

$$W(z) = \frac{1}{\beta_y(z)} \begin{pmatrix} \beta_\sigma^1(z) \\ \vdots \\ \beta_\sigma^p(z) \end{pmatrix},$$

i.e., the estimate $\bar{\sigma}$ of the functional σ satisfies the equation

$$\bar{\sigma} = \frac{1}{\beta_y(z)} \begin{pmatrix} \beta_\sigma^1(z) \\ \vdots \\ \beta_\sigma^p(z) \end{pmatrix} y, \tag{7.42}$$

and, as we can see from (7.42), the degree of the observer is equal to $\deg \beta_y(z)$. Among all observers satisfying conditions (7.41) we have to find an observer of the minimal order just as in the scalar case for which purpose we use the iteration procedure. To describe this procedure, we introduce notation

$$\varkappa = \deg \beta_y(z),$$

and this means that

$$CL = CAL = \dots = CA^{n-\varkappa-2}L = 0, \quad CA^{n-\varkappa-1}L \ne 0; \tag{7.43}$$

and, in addition, for every $i = 1, 2, \dots, p$ we have relations

$$F^i L = F^i AL = \dots = F^i A^{n-\varkappa-2} L = 0. \tag{7.44}$$

As in the preceding section, when solving systems of equations (7.43) and (7.44) we shall consider the original system in the canonical basis of observability. As before, we introduce vectors and matrices

$$L^\top = \begin{pmatrix} L^\varkappa \\ 0 \end{pmatrix} = (l_1, l_2, \dots, l_\varkappa, 1, 0, \dots, 0)^\top,$$

$$H_\varkappa^i = \begin{pmatrix} f_1^i & f_1^i & \vdots & f_\varkappa^i \\ f_2^i & f_3^i & \vdots & f_{\varkappa+1}^i \\ \vdots & \vdots & \vdots & \vdots \\ f_{n-\varkappa-1}^i & f_{n-\varkappa}^i & \vdots & f_{n-2}^i \end{pmatrix}, \quad h_\varkappa^i = \begin{pmatrix} f_{\varkappa+1}^i \\ f_{\varkappa+2}^i \\ \vdots \\ f_{n-1}^i \end{pmatrix}, \quad i = 1, 2, \dots, p,$$

where f_j^i $(j = 1, \dots, n)$ are components of the ith row F^i of the functional $\sigma = Fx$ in the canonical basis. We also define the matrix $H_\varkappa$ and the vector $h_\varkappa$ by the expressions

$$H_\varkappa = \begin{pmatrix} H_\varkappa^1 \\ H_\varkappa^2 \\ \vdots \\ H_\varkappa^p \end{pmatrix}, \quad h_\varkappa = \begin{pmatrix} h_\varkappa^1 \\ h_\varkappa^2 \\ \vdots \\ h_\varkappa^p \end{pmatrix}. \tag{7.45}$$

In these terms, the following statement is valid which generalizes Theorem 7.8 to the case of vector functional.

Theorem 7.10. *For the observable n-dimensional system*

$$x_{k+1} = Ax_{k+1}, \quad k = 0, 1, \dots$$

with a scalar output

$$y_k = Cx_k$$

and a linear p-dimensional vector functional

$$\sigma = Fx$$

such that rank $F = p$, rank $\binom{F}{C} = p + 1$, *and the pair $\{F, A\}$ is observable, there exists a functional observer of order $\varkappa$ of form* (7.42) *which reconstructs the functional σ if and only if in the canonical basis of observability the linear equation*

$$H_\varkappa l^\varkappa = -h_\varkappa \tag{7.46}$$

has a solution $l^\varkappa = (l_1, l_2, \dots, l_\varkappa)^\top$ such that $L^\varkappa = \binom{l^\varkappa}{1}$ is a Hurwitz vector.

The algorithm of computing the minimal $\varkappa$ described in the preceding section is also suitable for the case under consideration.

Remark 7.11. The requirement of observability of the pairs $\{C, A\}$ and $\{F, A\}$ can be lowered to detectability.

Scalar functional, vector observation. We shall restrict the consideration to the description of the main ideas of the synthesis of a functional observer in the present case and omit the details since they can be easily found from the material given above. Thus, we consider an n-dimensional system

$$x_{k+1} = Ax_k, \quad k = 0, 1, 2, \dots$$

with a vector l-dimensional output

$$y_k = Cx_k$$

which must be used for estimating the scalar functional

$$\sigma = Fx, \quad F \in \mathbb{R}^n.$$

We assume that rank $C = l$, the pair $\{C, A\}$ is observable, and ν is its observability index, i.e., ν is a minimal number such that

$$\operatorname{rank} \begin{pmatrix} C \\ CA \\ \vdots \\ CA^{\nu-1} \end{pmatrix} = n.$$

On these assumptions, by means of nonsingular transformations of the state and output

$$\bar{x} = Px, \quad \bar{y} = My, \quad \det P \neq 0, \quad \det M \neq 0,$$

the system under consideration can be reduced to the Luenberger–Isidori canonical form consisting of a set of l subsystems with scalar outputs y^i of the form

$$\begin{cases} x_{k+1}^i = A_{ii} x_k^i + \sum\limits_{j=1, i \neq j}^{l} a_{ij} y_k^j \\ y_k^i = c_i x_k^i, \quad k = 0, 1, 2, \dots, \end{cases} \tag{7.47}$$

where $i = 1, 2, \dots, l$; $x^i \in \mathbb{R}^{\nu_i}$, $\nu_1 + \nu_2 + \cdots + \nu_n = n$, $\nu = \max_i \nu_i$;

$$\bar{x} = \begin{pmatrix} x^1 \\ \vdots \\ x^l \end{pmatrix}, \quad \bar{y} = \begin{pmatrix} y^1 \\ \vdots \\ y^l \end{pmatrix}; \quad c_i = (0, \dots, 0, 1), \quad A_{ii} = \begin{pmatrix} 0 & \dots & 0 & * \\ 1 & \dots & 0 & * \\ \vdots & \ddots & \vdots & \vdots \\ 0 & \dots & 1 & * \end{pmatrix}.$$

Upon this transformation, in the new basis the functional $\sigma = Fx$ can be represented as the sum of l-functionals of the form $\sigma^i = F^i x^i$, i.e.,

$$\sigma = \sum_{i=1}^{l} F^i x^i = \sum_{i=1}^{l} \sigma^i. \tag{7.48}$$

It is easy to verify that if the pair $\{F, A\}$ is observable (and this fact will be assumed), then all pairs $\{F^i, A_{ii}\}_{i=1,\dots,l}$ will be similar.

This means that the problem under consideration can be reduced to a set l of subproblems of the estimation of scalar functionals

$$\sigma^i = F^i x^i$$

also with the use of scalar outputs

$$y_k^i = c_i x_k^i$$

of the systems

$$x^i_{k+1} = A_{ii} x_k, \quad k = 0, 1, 2, \dots .$$

Note that on the right-hand side of each ith equation there is a known input $\sum_{i=1, i \neq j}^{l} a_{ij} y^j_k$, but, as was repeatedly emphasized, this does not affect the synthesis of the kernel of the observer.

However, precisely with a problem of this kind we deal in the method of pseudoinputs, and its solution, as has been established, is an observer of the form

$$\bar{\sigma}^i = \frac{\beta^i_\sigma(z)}{\beta^i_y(z)} y^i, \tag{7.49}$$

where $\beta^i_\sigma(z)$ and $\beta^i_y(z)$ are numerators of transfer functions of the system

$$x^i_{k+1} = A_{ii} x^i_k + L^i v^i_k$$

from the pseudoinputs v^i_k acting along the vectors L^i to the outputs $\sigma^i = F^i x^i$ and $y^i = c^i x$ respectively, i.e.,

$$\bar{\sigma}^i = \frac{\beta^i_\sigma(z)}{\alpha^i(z)} v^i, \quad y^i = \frac{\beta^i_y(z)}{\alpha^i(z)} v^i,$$

where $\alpha^i(z) = \det(zI - A_{ii})$.

Thus, the general form of the required observer is given by the sum

$$\bar{\sigma} = \sum_{i=1}^{l} \bar{\sigma}^i = \sum_{i=1}^{l} \frac{\beta^i_\sigma(z)}{\beta^i_y(z)} y^i. \tag{7.50}$$

The order of observer (7.50) coincides with the order of the largest common multiple of the polynomials in the denominator, and, consequently, the observer has the minimal order when its zeros of polynomials $\beta^i_y(z)$ coincide (i.e., $\beta^i_y(z) = \beta_y(z)$ for all $i = 1, \dots, l$), are stable, and the degree of the polynomial $\beta_y(z)$ is minimal.

In this case, the final form of the required functional observer is defined by the operator expression

$$\beta_y(z)\sigma = \sum_{i=1}^{l} \beta^i_\sigma(z) y^i, \tag{7.51}$$

we should only take into consideration here that y^i are components of the transformed vector $\bar{y} = My$ where y are measurable variables of the output.

Observer (7.51) is physically realizable if

$$\deg \beta^i_\sigma(z) \leq \deg \beta_y(z), \quad i = 1, 2, \dots, l. \tag{7.52}$$

These inequalities and the conditions under which the polynomial $\beta_y(z)$ is a Hurwitz polynomial presuppose the conditions of choosing the vectors of the pseudoinputs L^i which, for $\varkappa = \deg \beta_y(z)$, must have the same structure

$$L^i = \begin{pmatrix} l_1 \\ \vdots \\ l_\varkappa \\ 1 \\ 0 \\ \vdots \\ 0 \end{pmatrix} \in \mathbb{R}^{\nu_i}, \quad L^\varkappa = \begin{pmatrix} l_1 \\ \vdots \\ l_\varkappa \\ 1 \end{pmatrix} \text{ is a Hurwitz vector,}$$

and the vector $l^\varkappa = (l_1, \dots, l_\varkappa)^{\top}$ is a solution of the set of linear equations

$$H^i_\varkappa l^\varkappa = -h^i_\varkappa, \quad i = 1, 2, \dots, l.$$

In this system of equations the matrix $H^i_\varkappa$ and the vector $h^i_\varkappa$ have the familiar form

$$H^i_\varkappa = \begin{pmatrix} f^i_1 & \dots & f^i_\varkappa \\ \vdots & \vdots & \vdots \\ f^i_{\nu_i-\varkappa-1} & \dots & f^i_{\nu_i-2} \end{pmatrix}, \quad h^i_\varkappa = \begin{pmatrix} f^i_{\varkappa+1} \\ \vdots \\ f^i_{\nu_i-1} \end{pmatrix}.$$

In order to calculate the minimal order of the observer, we should use the iteration procedure indicated above, beginning with the value $\varkappa^*$ for which the conditions

$$\operatorname{rank}(H^i_{\varkappa^*}) = \operatorname{rank}(H^i_{\varkappa^*}, h^i_{\varkappa^*})$$

are fulfilled.

7.4 Method of scalar observers in the problem of synthesis of a minimal order functional observer

We shall again consider a problem of synthesis of a minimal order observer which reconstructs a linear p-dimensional functional

$$\sigma = Fx,$$

from the measurements of the l-dimensional input

$$y_k = Cx_k, \quad k = 0, 1, 2, \dots,$$

of the n-dimensional stationary linear system

$$x_{k+l} = Ax_k, \quad k = 0, 1, 2, \dots .$$

We assume that F and C are full-rank matrices, i.e.,

$$\operatorname{rank} F = p, \quad \operatorname{rank} C = l,$$

the rank of the extended matrix $\binom{F}{C}$ being maximal, i.e.,

$$\operatorname{rank} \begin{pmatrix} F \\ C \end{pmatrix} = p + l \leq n,$$

in addition, the pairs of matrices $\{C, A\}$ and $\{F, A\}$ are observable.

On these assumptions it is clear that the problem of synthesis of the functional observer of minimal order $\varkappa$ is substantive if

$$p \leq \varkappa < n - l,$$

and, consequently, the strict inequality

$$p + l < n$$

should be valid.

We shall give an example where $p = l = 1$ to explain the idea of the method of scalar observers.

7.4.1 Scalar functional, scalar observation

Assume that the row of F satisfies the equation

$$FA = \lambda F + \mu C \tag{7.53}$$

for the real and stable number λ, i.e., $|\lambda| < 1$, and a certain μ. Then it is obvious that the asymptotic estimate $\bar{\sigma}$ of the functional $\sigma = Fx$ is given by a scalar observer of the form

$$\bar{\sigma}_{k+1} = \lambda \bar{\sigma}_k + \mu y_k, \quad k = 0, 1, 2, \ldots . \tag{7.54}$$

Since, in the general case, the row of F does not satisfy equation (7.53), the observation problem under consideration cannot be solved by a scalar observer of form (7.54). However, it can be solved by a set of scalar observers of form (7.54). Let us demonstrate this.

Suppose that the vector L provides the matrix

$$A_L = A - LC$$

with a real distinctive Hurwitz spectrum

$$\operatorname{spec}\{A_L\} = \{\lambda_1, \lambda_2, \ldots, \lambda_{(n-1)}, 0\},$$
$$\lambda_i \neq \lambda_j, \quad i \neq j; \quad |\lambda_i| < 1, \quad i = 1, 2, \ldots, n - 1.$$

We denote by g_i the left-hand eigenvectors of the matrix A_L associated with the eigenvalues λ_i. Then the set of vectors $g_1, \dots, g_{n-1}, C$ forms a basis and the row of F is uniquely decomposable according to this basis. Each functional $\xi_i = g_i x$ is reconstructed by a scalar observer similar to observer (7.54), i.e.,

$$\bar{\xi}^i_{k+1} = \lambda_i \bar{\xi}^i_k + \mu_i y_k, \quad \mu_i = g_i L, \tag{7.55}$$

and $y = Cx$ is a known output. Therefore we can state that the required functional σ can be reconstructed by the set of scalar observers (7.55), i.e.,

$$\bar{\sigma}_k = \sum_{i=1}^{n-1} w_i \bar{\xi}^i_k + w_n y_k, \quad k = 0, 1, 2, \dots, \tag{7.56}$$

and

$$\bar{\sigma}_k - \sigma_k = \sum_{i=1}^{n-1} w_i \varepsilon^i_k, \quad \varepsilon^i_k = \bar{\xi}^i_k - \xi^i.$$

It is precisely this circumstance that defines the name of the method[3].

In order to find the minimal number of scalar observers which can solve the problem under consideration, we must find a basis $g_1, \dots, g_{n-1}, C$ in $\mathbb{R}^n$ such that the vector $w = (w_1, \dots, w_n)$ in decomposition (7.56) would have the maximal number of zero components. This choice depends on the vector L (or the set $\{\lambda_1, \dots \lambda_{n-1}\}$), and, consequently, by using the requisite choice of the vector L we can solve the problem being considered.

Therefore the necessary and sufficient condition of existence of a functional observer of order $\varkappa$ is the condition of existence of a "Hurwitz" solution of the system of linear equations

$$H_\varkappa l^\varkappa = -h_\varkappa, \tag{7.57}$$

where

$$H_\varkappa = \begin{pmatrix} f_1 & f_2 & \dots & f_\varkappa \\ f_2 & f_3 & \dots & f_{\varkappa+1} \\ \vdots & \vdots & \vdots & \vdots \\ f_{n-\varkappa-1} & f_{n-\varkappa} & \dots & f_{n-2} \end{pmatrix}, \quad h_\varkappa = \begin{pmatrix} f_{\varkappa+1} \\ f_{\varkappa+2} \\ \vdots \\ f_{n-1} \end{pmatrix} \tag{7.58}$$

in the canonical basis of observability. Here f_i are the coordinates of the row of F in the indicated basis, i.e., $F = (f_1, f_2, \dots, f_n)$. In order to find the minimal $\varkappa^*$, we should use the iteration procedure described in the preceding subsections, with

$$\varkappa^* \le n - 1.$$

The given arguments sum up the analog of Theorem 7.8 whose formulation we omit.

[3] Note that the given arguments justify the Emelyanov–Taran hypothesis of the sixties about the possibility of replacement of the output differentiators by inertial links in the synthesis of feedback.

Remark 7.12.

1°. The synthesis of a family of scalar observers which reconstructs the given functional is carried out in the following sequence: after determining the minimal order of $\varkappa^*$ we find $l^\varkappa$, which is a solution of system (7.57), the component of the Hurwitz vector $L^\varkappa = \binom{l^\varkappa}{1}$, then we determine the diagonal $\varkappa \times \varkappa$ matrix $\Lambda^\varkappa = \mathrm{diag}(\lambda_1, \lambda_2, \ldots, \lambda_\varkappa)$, and find the eigenvectors g_i and, together with them, the estimates of the functionals $\bar{\xi}^i$. Finally, the required observer is defined by the relation

$$\begin{cases} \bar{\sigma}_{k+1} = w^\varkappa \bar{\xi}_k^\varkappa + w_n y_k \\ \bar{\xi}_{k+1}^\varkappa = \Lambda^\varkappa \bar{\xi}_k^\varkappa + \mu^\varkappa y_k, \quad k = 0, 1, 2, \ldots \end{cases} \tag{7.59}$$

where $w^\varkappa = (w_1, \ldots, w_\varkappa)$.

2°. In the case under consideration instead of the observability of the pairs $\{F, A\}$, $\{C, A\}$ we can only require their detectability.

3°. The spectrum $\mathrm{spec}\{A_L\} = \{\lambda_1, \lambda_2, \ldots, \lambda_{n-1}, 0\}$ can contain coincident roots or complex-conjugate pairs, and then the basis g_i will consist of eigenvectors and root vectors. We omit the details.

7.4.2 Scalar observations, vector functional

The technique of synthesis of minimal order functional observers described in the preceding subsections is preserved in the large, and therefore we shall only point out the main items. Thus, let rank $F = p > 1$, rank $\binom{F}{C} = p + 1$ and let $G \in \mathbb{R}^{(n-1)\times n}$ be a matrix consisting of left-hand eigenvectors g_i of the matrix A_L, i.e.,

$$A_L = \Lambda G.$$

The matrix F is uniquely "decomposed" according to the rows of the matrix G and the row of C, i.e.,

$$F = w'G + w_n C,$$

where $w' = (w_1, \ldots, w_{n-1})$, and therefore

$$\sigma = \begin{pmatrix} \sigma^1 \\ \vdots \\ \sigma^p \end{pmatrix} = w'\xi + w_n y, \quad \xi = Gx.$$

We denote $\Lambda = \mathrm{diag}(\lambda_1, \ldots, \lambda_{n-1})$. Since

$$\xi_{k+1} = \Lambda \xi_k + GLy_k, \quad k = 0, 1, 2, \ldots,$$

the generalized equation of the observer has the form

$$\begin{cases} \bar{\xi}_{k+1} = \Lambda \bar{\xi}_k + GLy_k \\ \bar{\sigma}_k = w'\bar{\xi}_k + w_n y_k \end{cases}$$

and now we should carry out the minimization, with respect to L, of its order (i.e., L is chosen such that among w_i the number of zeros should be maximal).

Let $\varkappa < n - 1$ be the order of the functional observer, and then the necessary and sufficient condition of solvability of the problem for the given $\varkappa$ is formulated in the form of existence of a "Hurwitz" solution $l^\varkappa$ of the systems of linear equations

$$H^i_\varkappa l^\varkappa = -h^i_\varkappa, \quad i = 1, 2, \ldots, p,$$

where $H^i_\varkappa$ and $h^i_\varkappa$ has the form indicated in (7.58) for each row F^i of the matrix F in the canonical basis of observability. In the large, the analog of Theorem 7.10 is valid, it has a strict formulation which we omit.

7.5 Synthesis of observers under the conditions of uncertainty

Let us consider a problem of observation of a phase vector in an n-dimensional linear stationary system

$$x_{k+1} = Ax_k + Df_k, \quad k = 0, 1, 2, \ldots, \tag{7.60}$$

from the measurements of its l-dimensional output

$$y_k = Cx_k. \tag{7.61}$$

Pay attention to the fact that the observed system is subjected to the action of the external signal f_k about which we know nothing *a priori* except of its dimension m, i.e., $f_k \in \mathbb{R}^m$.

Further, for full observability of the pair $\{C, A\}$, and in what follows we assume this property to be fulfilled, the standard full-dimensional observer

$$\bar{x}_{k+1} = A_k \bar{x}_k + Ly_k$$

does not solve the posed problem since the right-hand side of the equation for the estimation error $\varepsilon = \bar{x} - x$, i.e., the equation

$$\varepsilon_{k+1} = A_L \varepsilon_k - Df_k, \quad A_L = A - LC, \quad k = 0, 1, 2, \ldots,$$

includes, in general, the unknown disturbance $\{f_k\}$ which hinders the solution of the problem. However, under some additional conditions, it is possible to formulate the solution of this problem. These conditions will be formulated as the following assumptions:

(A.1) system (7.60), (7.61) is a hyperoutput system, i.e., $l > m$, or square when $l = m$,

(A.2) the matrices C, D, and CD are full-rank matrices, i.e., rank $C = l$, rank $D = m$, rank $CD = m$ $(l \geq m)$,

(A.3) the triplet $\{C, A, D\}$ is in the general position, i.e., the pair $\{C, A\}$ is observable and the pair $\{A, D\}$ is controllable,

(A.4) the triplet $\{C, A, D\}$ is of minimal phase, or, to put it otherwise, the invariant zeros of the matrix of the Rosenbrock system, i.e., the $(n+l) \times (n+m)$ matrices

$$R(z) = \begin{pmatrix} zI_n - A & | & -D \\ - & - & - \\ C & | & 0 \end{pmatrix},$$

are stable or absent.

Note that the indicated set of assumptions is standard in observation theory under uncertainty.

7.5.1 Square systems

We begin the consideration with the square systems (7.60) and (7.61) when $l = m$. In this case, there exists a nondegenerate change of coordinates M, $\det M \neq 0$,

$$\begin{pmatrix} x' \\ y \end{pmatrix} = Mx,$$

such that in the new coordinates $x' \in \mathbb{R}^{n-m}$, $y \in \mathbb{R}^m$ of the motion equation the system decomposes into two subsystems of equations

$$\begin{cases} x'_{k+1} = A_{11}x'_k + A_{12}y_k \\ y_{k+1} = A_{21}x'_k + A_{22}y_k + CDf_k, \quad k = 0, 1, 2, \ldots, \end{cases} \tag{7.62}$$

the first of which does not depend explicitly on the external disturbance f_k which affects only the second component whose state variable y_k can be immediately measured.

Thus, if the matrix A_{11} in (7.62) is stable, then the observation problem is solved by an observer of dimension $(n-m)$ of the form

$$\bar{x}'_{k+1} = A_{11}\bar{x}'_k + A_{12}y_k, \quad k = 0, 1, 2, \ldots, \tag{7.63}$$

with a static transformer

$$\bar{x}_k = M^{-1}\begin{pmatrix} \bar{x}'_k \\ y_k \end{pmatrix}. \tag{7.64}$$

However, as was already pointed out, the spectrum of the matrix A_{11} consists of the set of all invariant zeros of the Rosenbrock matrix of system $R(z)$ which, in turn, are stable if the triplet $\{C, A, D\}$ is of minimal phase (A.4). We have the following theorem.

Theorem 7.13. *Suppose that the square (i.e., $l = m$) minimal-phase triplet $\{C, A, D\}$ is in the general position,* rank $CD = m$. *Then the problem of observation of the full-phase vector of system* (7.60), (7.61), *with the presence of disturbance f_k, is solved by observer* (7.63) *and* (7.64) *of dimension $(n - m)$, the convergence of the estimate $\bar{x}_k$ to x_k being fully determined by the zeros of the Rosenbrock matrix $R(z)$ of the system.*

As concerns the synthesis of the functional observer (of the minimal order inclusive), here the possibilities are not very extensive and consist in the following.

The functional $\sigma = Fx$ being estimated is represented as the sum

$$\sigma = FM^{-1}\begin{pmatrix}\bar{x}'\\ y\end{pmatrix} = F'x' + F''y$$

whose second component is known, and therefore the posed problem reduces to the estimation of the functional

$$\sigma' = F'x', \quad F' \in \mathbb{R}^{1\times(n-m)}.$$

This is possible in principle. Thus, for instance, for the matrix A_{11} of simple structure in $\mathbb{R}^{n-m}$ there exists a basis consisting of its left-hand eigenvectors h^i $(i = 1, 2, \ldots, n - m)$ corresponding to its eigenvalues $\{\lambda_1, \lambda_2, \ldots, \lambda_{n-m}\}$, i.e.,

$$h^i A_{11} = \lambda_i h^i, \quad i = 1, 2, \ldots, n - m.$$

The vector F' is uniquely decomposed with respect to this basis

$$F' = \sum_{i=1}^{n-m} \mu_i h^i,$$

the number of nonzero factors of the decomposition defines the minimal order of the functional observer. By the known technique (see Chap. 5) this result can be generalized to the matrix A_{11} which is of an arbitrary structure. In any case, it is impossible to influence the order of the observer or the rate of convergence of the estimate to the original, the former and the latter are defined by the property of the original triplet $\{C, A, D\}$. The situation is different in the so-called hyperoutput systems where $l > m$.

7.5.2 Hyperoutput systems

In this case, when assumptions (A.1)–(A.4) are fulfilled in the equations of the system under consideration, we have to carry out the following transformations: we should begin with dividing the vector y into two components y' and y'' of dimension m and $(l - m)$

$$y = \begin{pmatrix}y'\\ y''\end{pmatrix} = \begin{pmatrix}C'x\\ C''x\end{pmatrix} \begin{matrix}\}\, m\\ \}\, l-m\end{matrix}$$

so that

$$\det C'D \neq 0.$$

Note that for this purpose we may require not only an interchange of rows in the matrix C', as was pointed out above, but also a nonsingular transformation of the output vector. However, this is a technical problem and does not affect the essence of the matter (for details see Chap. 5). After this transformation we have to consider a "square" system

$$\begin{cases} x_{k+1} = Ax_k + Df_k \\ y'_k = C'x_k, \end{cases}$$

for which we should carry out a change of variables indicated in the preceding item, namely,

$$x = M \begin{pmatrix} x' \\ y' \end{pmatrix} \begin{matrix} \}\, m \\ \}\, l - m \end{matrix}$$

and obtain a system

$$\begin{cases} x'_{k+1} = A_{11}x'_k + A_{12}y'_k \\ y'_{k+1} = A_{21}x'_k + A_{22}y'_k + (C'D)f_k, \quad k = 0, 1, 2, \dots . \end{cases} \tag{7.65}$$

The significant difference between systems (7.65) and (7.62) is that in the case of systems (7.65) the first subsystem is followed by the output

$$y'' = C''x = C''M \begin{pmatrix} x' \\ y' \end{pmatrix} = C''_1 x' + C''_2 y \tag{7.66}$$

in whose equation the second component is known, and therefore, when solving observation problems with the presence of disturbance, we may deal with an $(n - m)$-dimensional system with an output of dimension $(l - m) > 0$, i.e., a system of the form

$$\begin{cases} x'_{k+1} = A_{11}x'_k + A_{12}y'_k \\ \bar{y}_k = C''_1 x'_k, \quad k = 0, 1, 2, \dots . \end{cases} \tag{7.67}$$

Of fundamental importance for the further use of system (7.67) in the estimation problem under consideration is the question concerning the generality of position of the pair $\{C''_1, A_{11}\}$. This question is answered by the following lemma.

Lemma 7.14. *Suppose that the minimal-phase triplet $\{C, A, D\}$ is in the general position,* rank $CD = m$. *Then there exist nonsingular transformations of the output and state of the system such that in the transformed system* (7.63) *and* (7.64) *the pair $\{C''_1, A_{11}\}$ is observable* (*reconstructible*) *when the Rosenbrock matrix of the system does not have invariant zeros* (*there are stable invariant zeros*).

We can prove Lemma 7.14 with the use of the arguments employed in the proof of Theorem 5.1 in Chap. 5.

Thus, the full-dimensional observer for system (7.67) does not contain uncertainty, can be taken in standard, for the theory of observability, form of an $(n-m)$-dimensional system

$$\bar{x}'_{k+1} = A^L_{11}\bar{x}'_k + L\bar{y}_k + A_{12}y'_k, \tag{7.68}$$

where the matrix

$$A^L_{11} = A_{11} - LC''_1. \tag{7.69}$$

Under the conditions of Lemma 7.14 the spectrum of the matrix A^L_{11} is a Hurwitz spectrum and contains two components, namely, an unchangeable component consisting of invariant zeros of the Rosenbrock matrix $R(z)$ and a variable component defined arbitrarily by the choice of the vector L in (7.69). Observer (7.68) together with the transformation[4]

$$\bar{x}_k = M\begin{pmatrix}\bar{x}'_k \\ y'_k\end{pmatrix} \tag{7.70}$$

solves the posed problem of observation under uncertainty. We have thus proved the following theorem.

Theorem 7.15. *For the minimal-phase hyperoutput* $(l > m)$ *triplet* $\{C, A, D\}$ *of the general position, under the condition* $\operatorname{rank} CD = m$, *the problem of observation of the full-phase vector, at the presence of the unknown disturbance* $\{f_k\}$, *is solved by observer* (7.68)–(7.70) *of order* $(n-m)$. *In this case, the asymptotics of estimation is defined by the matrix* $A^L_{11} = A_{11} - LC''_1$ *whose spectrum is formed by all invariant zeros of the Rosenbrock matrix* $R(z)$ *of the system and by the elements defined arbitrarily by the requisite choice of the vector* $L \subset \mathbb{R}^{(n-m)\times(l-m)}$.

Thus, comparing Theorems 7.13 and 7.15, we can infer that the presence of additional outputs allows us to affect essentially the dynamics of the system of estimation, and, in certain situations, we can define it arbitrarily.

In addition to full-dimensional (of order $(n-m)$) observer described above for estimating the phase vector of system (7.67), we can also use a lowered order observer of order $(n-l)$. For this purpose we must use Lemma 7.6 from Sec. 7.2 according to which by a nonsingular transformation equations (7.67) are reduced to a system of equations of form

$$\begin{cases} x''_{k+1} = \bar{A}_{11}x''_k + \bar{A}_{12}\bar{y}_k + A'_{12}y'_k \\ \bar{y}_{k+1} = \bar{A}_{21}x''_k + \bar{A}_{22}\bar{y}_k + A''_{12}y'_k, \quad k = 0, 1, 2, \dots, \end{cases} \tag{7.71}$$

where the first equation is of dimension $(n-l)$ with matrix $\bar{A}_{11}$ with a spectrum consisting of all invariant zeros of the Rosenbrock matrix $R(z)$ of the system and

[4]Note that if we use the output transformation when solving equations (7.65), we have to change in (7.70) the "transformed" component y'_k by its preimage.

the elements of this spectrum defined arbitrarily. In (7.71) (A'_{12}) and A''_{12} are matrices obtained from the matrix A_{12} as a result of the change of coordinates mentioned above.

Now, as before, an observer of form

$$\bar{x}''_{k+1} = \bar{A}_{11}\bar{x}''_k + \bar{A}_{12}\bar{y}_k + A'_{12}y'_k, \quad k = 0, 1, 2, \dots, \tag{7.72}$$

is suitable for estimating the vector $x''_k \in \mathbb{R}^{n-l}$. This observer, together with the relation

$$\bar{x}_k = \bar{M}\begin{pmatrix} \bar{x}''_k \\ y_k \end{pmatrix}, \tag{7.73}$$

for a certain nonsingular matrix $\bar{M}$ gives the required estimate of the full-phase vector of the original system. We shall sum up what has been said as the following statement.

Theorem 7.16. *For the minimal-phase hyperoutput $(l > m)$ triplet $\{C, A, D\}$ of the general position, under the additional condition* rank $CD = m$, *the problem of observation of the full-phase vector with the presence of the unknown disturbance $\{f_k\}$ is solved by observer* (7.72), (7.73) *of the minimal possible order $(n - l)$. In this case, the dynamics of estimation is defined by all invariant zeros of the Rosenbrock matrix $R(z)$ of the system and the other points of the spectrum are defined arbitrarily.*

For full-dimensional observers the result established in Theorem 7.16 is the most possible from the point of view of minimization of dynamic order of the observer. The further lowering of the order is possible only for functional observers.

As applied to system (7.67), the synthesis of the observer which estimates the functional[5]

$$\sigma' = F'x', \quad x' \in \mathbb{R}^{n-l},$$

reduces to one of the sequences of actions described above. For instance, for the observable pair $\{C'_1, A_{11}\}$ in the case where $C'_1 \in \mathbb{R}^{1\times l}$, in accordance with the methods of scalar observers, we should find the matrix H' of the left-hand correspondingly vectors of the matrix $A^L_{11} = A_{11} - LC''_1$, i.e.,

$$H'A^L_{11} = \Lambda H',$$

where $\Lambda = \operatorname{diag}(\lambda_1, \lambda_2, \dots, \lambda_{n-l-1})$ is a matrix of eigenvalues of the matrix A^L_{11}. Then we should find the parameters $w = (w_1, \dots, w_{n-l-1}, w_{n-l}) = (w', w_{n-l})$ of the decomposition

$$F' = w'H' + w_{n-l}C''_1$$

and reduce the problem of vector estimation to a set of scalar problems of estimation

$$\sigma'_i = h^i x' \quad (i = 1, 2, \dots, n - l - 1),$$

[5] As was pointed out above, the estimation of the functional $\sigma = Fx$ by a change $x = \bar{M}\begin{pmatrix} \bar{x}' \\ y \end{pmatrix}$ reduces to the estimation of the functional $\sigma' = F'x'$.

where h^i are rows of the matrix H^i corresponding to the respective value λ_i. Then the required estimate will be given by the sum of estimates

$$\bar{\sigma}' = \sum_{i=1}^{n-l-1} w_i \bar{\sigma}_i' + w_{n-l} y, \tag{7.74}$$

where w_i are components of the row w and $\bar{\sigma}_i'$ is the estimate of the functional $\sigma_i' = h^i x'$ formed by the scalar observer

$$(\bar{\sigma}_i')_{k+1} = \lambda_i (\bar{\sigma}_i')_k + d_i y_k, \quad k = 0, 1, 2, \dots, \tag{7.75}$$

where d_i is a row by which all outputs y_k of the original system are acting. The minimization of the number of components in decomposition (7.74) can be achieved with the use of the iteration process. This minimization is reduced to finding the minimal number $\varkappa$ for which the system of linear equations

$$H_\varkappa l^\varkappa = -h_\varkappa \tag{7.76}$$

has, as its solution, a vector $l^\varkappa$ which is a Hurwitz component of the vector $L^\varkappa = \binom{l^\varkappa}{1}$. In (7.76) the matrices $H_\varkappa$ and the vector $h_\varkappa$ consist of the coordinates of the vector F' in the canonical basis of observability of the pair $\{C_1'', A_{11}\}$. We omit the details (see Chap. 4). As before, in similar cases we shall indicate that the requirement of observability of the pair $\{C_1'', A_{11}\}$ can be weakened to the property of reconstructibility.

7.5.3 Method of pseudoinputs in the synthesis of state observer

The approach described in the preceding item, where, for obtaining a full-dimensional state observer of minimal order, we require, in general, two arbitrary successive similarity transformations, has an alternative which is the method of pseudoinputs where we can use only one transformation of this kind.

Let us consider again the hyperoutput (i.e., $l > m$) system (7.60) and (7.61) and carry out its "squarefication" by adding $(l - m)$ new zero inputs acting on the system via the matrix D' so that the system is defined by the equations

$$\begin{cases} x_{k+1} = A x_k + D f_k + D' f_k' = A x_k + \bar{D} \bar{f}_k, & k = 0, 1, 2, \dots \\ y_k = C x_k. \end{cases} \tag{7.77}$$

It stands to reason that it is the preceding system since $f_k' \equiv 0$ but in this case it is immersed into the class of square $(l \times l)$ systems. The important difference of the transformed system is that the Rosenbrock matrix of the extended system

$$R'(z) = \begin{pmatrix} zI_n - A & | & D : D' \\ -- & -- & -- \\ C & | & 0 \end{pmatrix} \tag{7.78}$$

has new invariant zeros which depend on the matrix D' and which are also zeros of the characteristic polynomial of zero dynamics

$$\beta'(z) = \det R'(z), \quad \deg \beta'(z) = n - l. \tag{7.79}$$

In greater detail this is stated in the following lemma.

Lemma 7.17. *Suppose that the minimal-phase triplet $\{C, A, D\}$ is in the general position,* rank $CD = m$. *Then the set of $(n - l)$ zeros of the characteristic polynomial of zero dynamics of the extended system $\beta'(z)$ consists of the set of invariant zeros of the Rosenbrock matrix $R(z)$ of the original system and a set of zeros defined arbitrarily by a requisite choice of the pseudoinput matrix D', in this case* $\operatorname{rank}(C(D\,;\,D')) = l$.

We omit the proof since it can be reconstructed with the use of arguments from Chap. 5, see the proof of Theorem 5.3.

Proceeding from Lemma 7.17, we can realize a single nondegenerate transformation

$$x = M \begin{pmatrix} x' \\ y \end{pmatrix}$$

of the original equation (7.60) (or, what is the same, of the extended equation (7.77)) in order to reduce it to the form similar to the set of equations (7.65) and (7.71), namely, to a system of equations of the form

$$\begin{cases} x'_{k+1} = A_{11}x'_k + A_{12}y_k \\ y_{k+1} = A_{21}x'_k + A_{22}y_k + C(D\,;\,D')\bar{f}_k, \quad k = 0, 1, 2, \dots, \end{cases}$$

which, with due account of the fact that $(D\,;\,D')\bar{f}_k = Df_k$, finally assumes the form

$$\begin{cases} x'_{k+1} = A_{11}x'_k + A_{12}y_k \\ y_{k+1} = A_{21}x'_k + A_{22}y_k + CDf_k, \quad k = 0, 1, 2, \dots, \end{cases} \tag{7.80}$$

in which $x' \in \mathbb{R}^{n-l}$ and the spectrum of the matrix A_{11} coincides with the set of zeros of the polynomial $\beta'(z)$ in (7.79), and if A_{11} is a Hurwitz matrix, then the posed problem of estimation of the full-phase vector of system (7.60), (7.61), with the presence of an unknown disturbance, is solved by an observer of the minimal possible order $(n - l)$

$$\bar{x}'_{k+1} = A_{11}\bar{x}'_k + A_{12}y_k, \quad k = 0, 1, 2, \dots, \tag{7.81}$$

and a static transformer

$$\bar{x}_k = M \begin{pmatrix} \bar{x}'_k \\ y_k \end{pmatrix}, \tag{7.82}$$

where M is the nonsingular transformation, mentioned above, of the original system to form (7.80).

We have thus established the following theorem.

Theorem 7.18. *For the minimal-phase hyperoutput* ($l > m$) *triplet* $\{C, A, D\}$ *of the general position, under the additional condition* $\operatorname{rank} CD = m$, *the problem of observation of the full-phase vector with the presence of the unknown disturbance* $\{f_k\}_0^\infty$ *is solved by observer* (7.81), (7.82) *of the minimal possible order* $(n - l)$.

The following two circumstances may serve as a comment to this theory.

1°. Functional observers, of the minimal order inclusive, for system (7.60) and (7.61) must be defined by the first equations from (7.80) and use the iteration procedures described above.

2°. We should not always tend to the minimal dimension of the observer "cleaned" of disturbance, we can use any dimension from $(n - m)$ to $(n - l)$ inclusive. The following theorem gives the theoretical basis for this.

Theorem 7.19. *Let* ρ *be any number such that* $0 \le \rho \le l - m$. *Then, for any hyperoutput* ($l > m$) *minimal-phase triplet* $\{C, A, D\}$, *which is in the general position and satisfies the condition* $\operatorname{rank} CD = m$, *there exist nonsingular transformations of the vectors of the system and its phase vector such that the system can be reduced to the set of equations*

$$\begin{cases} (x'_\rho)_{k+1} = A_{11}(x'_\rho)_k + A_{12}(y_\rho)_k \\ (y_\rho)_{k+1} = A_{21}(x'_\rho)_k + A_{22}(y_\rho)_k + C_\rho D f_k \\ (y_{l-m-\rho})_k = C_{l-m-\rho}(x'_\rho)_k, \quad k = 0, 1, 2, \dots, \end{cases} \tag{7.83}$$

where $x'_\rho \in \mathbb{R}^{n-m-\rho}$; $y_\rho \in \mathbb{R}^{m+\rho}$; $y^{l-m-\rho} \in \mathbb{R}^{l-m-\rho}$ *is a part of the transformed output of the system,* $y = P_\rho \begin{pmatrix} y_\rho \\ y_{l-m-\rho} \end{pmatrix}$; $\operatorname{rank}(C_\rho D) = m$; A_{11} *is a Hurwitz matrix, for* $\rho > 0$ *its spectrum is partially defined arbitrarily.*

In addition, the observer which solves the problem of estimation of the full-phase vector of the system, is defined by the equations

$$\begin{cases} (\bar{x}'_\rho)_{k+1} = A_{11}(\bar{x}'_\rho)_k + A_{12}(y_\rho)_k \\ \bar{x}_k = M_\rho \begin{pmatrix} (\bar{x}'_\rho)_k \\ (y_\rho)_k \end{pmatrix}, \quad k = 0, 1, 2, \dots, \end{cases} \tag{7.84}$$

where M_ρ is the matrix of the transformation of the phase vector mentioned above.

Note that Theorem 7.19 also generalizes Theorems 7.16 and 7.18 in that part where the pair $\{C_{l-m-\rho}, A_{11}\}$ is reconstructible or observable depending on the presence or absence of stable invariant zeros of the Rosenbrock matrix $R(z)$ of the system.

7.5.4 Some classical methods of synthesis of state observers under the conditions of uncertainty

We shall briefly describe the classical schemes of synthesis of state observers under the conditions of uncertainty restricting the discussion on the exposition (in contrast to the results considered in Chap. 5 for systems of continuous time) of the results for square systems, i.e., systems of the form

$$\begin{cases} x_{k+1} = Ax_k + Df_k \\ y_k = Cx_k, \quad k = 0, 1, 2, \dots, \end{cases} \tag{7.85}$$

where $x \in \mathbb{R}^n$, f and $g \in \mathbb{R}^m$. These results can be generalized to hyperoutput systems ($l > m$) with the use of the arguments given in Chap. 5 or in original works.

In the sequel, without loss of generality, we shall assume that for the square triplet $\{C, A, D\}$ assumptions (A.1)–(A.4) are fulfilled, i.e., the triplet is of minimal phase and of general position and $\det(CD) \neq 0$.

7.5.5 Method of exclusion of perturbation from the equation for the estimation error

We choose an observer in the form of an n-dimensional dynamical system

$$w_{k+1} = Ew_k + Qy_k, \quad k = 0, 1, 2, \dots, \tag{7.86}$$

and the static transformer

$$\bar{x}_k = w_k + Ly_k. \tag{7.87}$$

In this case, the motion equation for the estimation error

$$\varepsilon = \bar{x} - x$$

has the form

$$\varepsilon_{k+1} = E\varepsilon_k + (EP + QC - PA)x_k - PDf_k, \tag{7.88}$$

where the matrix $P = I_n - LC$. The right-hand side of equation (7.88) does not depend on the unknown functions x_k and f_k when the relations

$$\begin{aligned} PD &= 0 \\ PA &= EP + QC \end{aligned} \tag{7.89}$$

are satisfied. If, in addition, E is a Hurwitz matrix, then the observer defined by equations (7.86) and (7.87) solves the observation problem under consideration since the equation

$$\varepsilon_{k+1} = E\varepsilon_k, \quad k = 0, 1, 2, \dots,$$

is asymptotically stable. Let us consider conditions (7.89) in greater detail. It follows from the relation $PD = 0$ that

$$P = I_n - D(CD)^{-1}C, \tag{7.90}$$

i.e., P is a matrix of the operator of the nonorthogonal projection "along C onto $Dx = 0$". Postmultiplying now the second relation in (7.89) by the matrix D, we obtain an equation for the matrix Q:

$$PAD = QCD$$

whose solution is the $n \times m$ matrix

$$Q = PAD(CD)^{-1}.$$

With due account of the last relation, the second relation in (7.89) assumes the form

$$PAP = EP, \tag{7.91}$$

which implies a relation

$$CEP = 0.$$

This means that as the matrix E we can take any matrix satisfying the relation

$$CE = \Lambda_1 C, \tag{7.92}$$

where Λ_1 is a diagonal $(m \times m)$ matrix.

Let us again consider equation (7.91) and assume that the $(n - m) \times n$ matrix H is a matrix of other, differing from C, left-hand eigenvectors of the matrix E with a diagonal matrix of the eigenvalues Λ_2, i.e.,

$$HE = \Lambda_2 H,$$

so that $\begin{pmatrix} C \\ H \end{pmatrix} E = \begin{pmatrix} \Lambda_1 & 0 \\ 0 & \Lambda_2 \end{pmatrix} \begin{pmatrix} C \\ H \end{pmatrix}$.

Then HP is a matrix of left-hand eigenvectors of the matrix AP since

$$(HP)AP = HEP = \Lambda_2(HP).$$

This means that Λ_2 is a matrix of nonzero eigenvalues of the matrix AP, but the latter is the spectrum of zero dynamics or, what is the same, invariant zeros of the Rosenbrock matrix $R(z)$ of the system.

We have thus proved the following statement.

Statement 7.20. Suppose that the square minimal-phase triplet $\{C, A, D\}$ is in the general position and $\det CD \neq 0$. Then there exists a Hurwitz $(n \times n)$ matrix E whose spectrum is composed of the whole spectrum of zero dynamics of the system and the remaining part is defined arbitrarily and is such that the problem of estimation of the full-phase vector of the system at the presence of unknown disturbance is solved by observer (7.86), (7.87).

Remarks. 1. A similar result can be formulated for the hyperoutput system ($l > m$) when rank $CD = m$.
2. The dimension of the observer can be lowered to $(n - m)$ when the dynamics of the estimator is fully defined by a spectrum of zero dynamics.
3. Equations (7.86) can be used for synthesizing functional observers, we omit the details.

7.5.6 Method of exclusion of perturbation from the equation of the system

We shall again consider, under conditions (A.1)–(A.4), the square $(m \times m)$-system

$$\begin{cases} x_{k+1} = Ax_k + Df_k \\ \quad y_k = Cx_k, \quad k = 0, 1, 2, \ldots, \end{cases}$$

whose first equation is transformed by means of the nonorthogonal projection mentioned in the preceding item, i.e., we introduce a new variable

$$\xi = Px, \quad P = I_n - D(CD)^{-1}C, \tag{7.93}$$

and then we obtain an equation (since $PD = 0$)

$$\xi_{k+1} = PA\xi_k + PAD(CD)^{-1}y_k \tag{7.94}$$

whose right-hand side does not depend explicitly on the external disturbance f_k. Note that the employed change (7.93) is not invertible. Moreover, we have an obvious relation $C\xi_k = 0$, $k = 0, 1, 2, \ldots$. This means that (7.94) contains only $(n - m)$ linearly independent rows and, consequently, the number of equations in (7.94) can be made smaller. For instance, this can be done in the following manner. Assuming that in the decomposition

$$C = (C_{n-m}\,;\, C_m)$$

the matrix C_m is invertible, we express, in the equation $C\xi = 0$, the last m-components of the vector ξ in terms of its first $(n - m)$ components, i.e.,

$$\xi_k^m = C_m^{-1}C_{n-m}\xi_k^{n-m}. \tag{7.95}$$

Then, in equation (7.94) we remove the last m rows and in the other rows make a change (7.95). As a result we obtain an equation of order $(n - m)$

$$\xi_{k+1}^{n-m} = A_{11}\xi_k^{n-m} + A_{12}y_k, \quad k = 0, 1, 2, \ldots . \tag{7.96}$$

Note that y_k cannot be expressed by ξ_k since

$$y_k = Cx_k = C(\xi_k + D(CD)^{-1}y_k) = y_k.$$

Therefore, for obtaining full motion equations, we must complement (7.96) by an equation for y_{k+1}, i.e., equation

$$y_{k+1} = C(A_{11}x_k + Df_k)|_{x=\xi+D(CD)^{-1}y} = A_{21}\xi_k^{n-m} + A_{22}y_k + CDf_k. \quad (7.97)$$

For the minimal-phase triplet $\{C, A, D\}$ the matrix A_{11} is a Hurwitz matrix, and therefore the observer is the dynamical system

$$\bar{\xi}_{k+1}^{n-m} = A_{11}\bar{\xi}_k^{n-m} + A_{12}y_k, \quad k = 0, 1, 2, \dots, \quad (7.98)$$

and the static transformer

$$\bar{x}_k = \begin{pmatrix} E_{n-m} \\ C_m^{-1}C_{n-m} \end{pmatrix} \bar{\xi}_k^{n-m} + D(CD)^{-1}y_k, \quad (7.99)$$

and this solves the posed problem of estimation of the phase vector under uncertainty. Moreover, we have the following statement.

Statement 7.21. For the square minimal-phase triplet $\{C, A, D\}$ of the general position such that $\det CD \neq 0$, the problem of estimation of the full-phase vector of the system with an unknown disturbance $\{f_k\}$ is solved by observer (7.98) and (7.99) of order $(n - m)$. The convergence of the estimate is fully defined by the invariant zeros of the Rosenbrock matrix $R(z)$.

Remarks. 1. The described transformation method is known as the *method of quasi-splitting* and was proposed in 1984 by S. K. Korovin. It is described, for instance, in [6].
2. It stands to reason that this method can be generalized to hyperoutput systems when, instead of (7.96) and (7.97), we have to deal with equations of the form

$$\begin{cases} \xi_{k+1}^{n-m} = A_{11}\xi_k^{n-m} + A_{12}y_k \\ y'_{k+1} = A_{21}\xi_k^{n-m} + A_{22}y'_k + (C'D)f_k \\ y''_k = C''\xi_k^{n-m}, \quad k = 0, 1, 2, \dots, \end{cases} \quad (7.100)$$

where C' is an m-row matrix such that $\det(C'D) \neq 0$ and C'' is an $(l - m)$-row matrix of the shortened output y''.

It is significant here that the pair $\{C'', A_{11}\}$ is observable when the Rosenbrock matrix $R(z)$ of the system does not have invariant zeros and is reconstructible when it has these zeros which, of course, under conditions (A.1)–(A.4), are stable.

Therefore, the equations of the observer which solves the problem under consideration can be taken in the form

$$\begin{cases} \bar{\xi}_{k+1}^{n-m} = A_{11}^L\bar{\xi}_k^{n-m} + \bar{A}_{12}y_k \\ \bar{x}_k = H\bar{\xi}_k^{n-m} + Ny_k \end{cases} \quad (7.101)$$

for certain matrices H and N. Here $A_{11}^L = A_{11} - LC''$ is a stable matrix whose spectrum is formed by all invariant zeros of the system and the remaining zeros are defined arbitrarily. In this case the full analog of Statement 7.21 is valid whose formulation is omitted here.
3. The dimension of an observer of form (7.101) can be lowered to $(n - l)$.
4. For synthesizing functional observers we can use equations (7.100) or their analogs.

7.5.7 Methods based on special canonical forms

The methods belonging to this group are based on two successive transformations which give, as a result, equations similar to equations (7.100).

We begin with finding an invertible $(n \times n)$ matrix T such that

$$TD = \begin{pmatrix} 0 \\ \bar{D} \end{pmatrix} \begin{matrix} \}\, n - m \\ \}\, m \end{matrix}$$

and carry out a change of variables

$$Tx = \begin{pmatrix} x' \\ x'' \end{pmatrix} \begin{matrix} \}\, n - m \\ \}\, m \end{matrix}$$

which would make it possible to reduce the original system (7.60) and (7.61) under conditions (A.1)–(A.4) to the form

$$\begin{cases} x'_{k+1} = A_1 x'_k + A_2 x''_k \\ x''_{k+1} = A_3 x'_k + A_4 x''_k + \bar{D} f_k \\ y_k = C_1 x'_k + C_2 x''_k, \quad k = 0, 1, 2, \dots, \end{cases} \tag{7.102}$$

where $\det C_2 \neq 0$. Removing from (7.102) the second recurrence equation and expressing in the first equation x'' by y and x', we can obtain standard equations which do not contain, in explicit form, the unknown disturbance $\{f_k\}$, i.e., equations

$$\begin{cases} x'_{k+1} = A_{11} x'_k + A_{12} y_k \\ y'_k = \bar{C}_1 x'_k, \quad k = 0, 1, 2, \dots, \end{cases} \tag{7.103}$$

where the pair $\{\bar{C}_1, A_{11}\}$ is observable when the original system does not have invariant zeros and only reconstructible otherwise (recall that invariant zeros are stable according to assumptions (A.1)–(A.4)). This means that the required observer may be the dynamical system of order $(n - m)$ of the form

$$\begin{cases} \bar{x}'_{k+1} = A_{11}^L \bar{x}'_k + \bar{A}_{12} y_k \\ \bar{x}_k = H \bar{x}'_k + N y_k, \quad k = 0, 1, 2, \dots, \end{cases} \tag{7.104}$$

where the matrix $A_{11}^L = A_{11} - L\bar{C}_1$ has a Hurwitz spectrum with ordinary properties. Moreover, we have the following statement.

Statement 7.22. For the hyperoutput ($l > m$) minimal-phase triplet $\{C, A, D\}$ of the general position such that $\det CD = m$, the problem of estimation of the full-phase vector of the system at the presence of the unknown disturbance $\{f_k\}$ is solved by an observer of form (7.104) of order $(n - m)$. The convergence of the estimate to the original is defined by the matrix A_{11}^{L} whose spectrum is formed by all invariant zeros of the Rosenbrock matrix $R(z)$ of the system and the values defined arbitrarily.

Remarks.

1. The order of the observer can be lowered to $(n - l)$ by a standard manner.

2. For synthesizing functional observers under uncertainty we can use equations (7.103) and some other observers of this kind, we omit the details.

3. We can see that the methods described above can be used under the same conditions and are based on similar ideas. The main difference between them is in the computation methods which we have to use for solving problems of the synthesis of observers.

Conclusion

In Chap. 7 we considered the methods of synthesis of observers for discrete systems. The presented results are similar to the corresponding results described in Chaps. 2–5 for continuous systems.

In Sec. 7.1 we gave general information concerning the theory of discrete dynamical systems.

In Sec. 7.2 we considered the concepts of observability and reconstructibility for discrete systems and gave criteria of observability and canonical forms for discrete systems.

In Sec. 7.3 we exposed the methods of synthesis of observers of a full-phase vector for discrete systems.

In Sec. 7.4 we considered the problem of synthesis of functional observers and in Sec. 7.5 the problem of synthesis of observers under uncertainty.

Bibliography

[1] Andreev, Yu.N. (1976): Control of Linear Finite-dimensional Objects (in Russian). Nauka, Fizmatlit, Moscow.

[2] Barbashin, E.A. (1967): Introduction to the Theory of Stability (in Russian). Nauka, Fizmatlit, Moscow.

[3] Voronov, A.A. (1976): Stability, Controllability, Observability (in Russian). Nauka, Fizmatlit, Moscow.

[4] Gantmacher, F.R. (2004): Theory of Matrices (in Russian). Nauka, Fizmatlit, Moscow.

[5] Emelyanov, S.V., Korovin, S.K., Nersisyan, A.L. (1990): On asymptotic properties of state observers for uncertain systems with a linear stationary part. Dokl. Akad. Nauk SSSR, **311**, 4, 807–811.

[6] Emelyanov, S.V., Korovin, S.K. (1997): New Types of Feedback. Control under Uncertainty (in Russian). Nauka, Fizmatlit, Moscow.

[7] Emelyanov, S.V., Korovin, S.K. (2003): Collection "Mathmatical Modelling. Problems and Results" (in Russian). Nauka, Fizmatlit, 12–35.

[8] Emelyanov, S.V., Korovin, S.K., Shepitko, A.S. (2000): Stahillzation of bilinear systems in the plane by means of constant and relay controls. Dif. Uravn., **36**, 8, 1021–1028.

[9] Ilyin, A.V., Korovin, S.K., Fomichev, V.V. (1997): Algorithms of inversion of linear controlled systems. Dif. Uravn., **34**, 6, 744–750.

[10] Ilyin, A.V., Korovin, S.K., Fomichev, V.V. (1998): Robust inversion of vector systems. Dif. Uravn., **34**, 11, 1478–1486.

[11] Ilyin, A.V., Korovin, S.K., Fomichev, V.V. (1997): Algorithms of inversion of linear scalar dynamical systems: method of controlled model. Dif. Uravn., **33**, 3, 329–339.

[12] Ilyin, A.V., Korovin, S.K., Fomichev, V.V. (2004): Algorithms of robust inversion of vector linear systems, in: Nonlinear Dynamics and Control, issue 4. Collection of Articles, S. V. Emelyanov, S. K. Korovin, eds. (in Russian). Fizmatlit, Moscow, 17–22.

[13] Ilyin, A.V., Korovin, S.K., Fomichev, V.V., Khlavenka, A. (2005): Synthesis of asymptotic observers for linear vector uncertain systems. Dif. Uravn., **41**, 1, 73–81.

[14] Ilyin, A.V., Korovin, S.K., Fomichev. V.V. (2005): Asymptotic observers with discontinuous control for scalar linear uncertain systems. Dif. Uravn., **41**, 10, 1310–1317.

[15] Ilyin, A.V., Korovin, S.K., Fomichev, V.V., Khlavenka, A. (2005): Dif. Uravn., **41**, 11, 1443–1457.

[16] Korovin, S.K., Medvedev, I.S., Fomichev, V.V. (2004): Functional observers for linear systems with uncertainty. Dif. Uravn., **42**, 10, 1307–1317.

[17] Ilyin, A.V., Korovin, S.K., Fomichev, V.V. (2006): On equations and properties of zero dynamics of linear controlled stationary systems. Dif. Uravn., **42**, 12, 1626–1636.

[18] Korovin, S.K. Ilyin, A.V., Fomichev, V.V. (1997): Method of controllable model in problems of inversion of dynamical systems. Dokl. Ros. Akad. Nauk, Control Theory, **354**, 2, 171–173.

[19] Korovin, S.K., Ilyin, A.V., Fomichev, V.V. (1998): Robust inversion of controlled linear systems. Dokl. Ros. Akad. Nauk, Control Theory, **356**, 2, 121–123.

[20] Korovin, S.K. Ilyin, A.V., Fomichev, V.V., Khlavenka, A. (2004): Asymptotic observers of the state of uncertain vector linear systems. Dokl. Ros. Akad. Nauk, Control Theory, **396**, 4, 469–473.

[21] Korovin, S.K., Fomichev, V.V. (2001): Asymptotic observers in the plane. Dif. Uravn., **37**, 12.

[22] Korovin, S.K., Fomichev, V.V. (2001): Exponential observers of bilinear systems in the plane. Dokl. Ros. Akad. Nauk, Control Theory, **385**, 45, 713–728.

[23] Korovin, S.K., Fomichev, V.V. (2002): Construction of exponential observers for bilinear controllable systems. Dif. Uravn., **38**, 1, 139–140.

[24] Korovin, S.K., Fomichev, V.V. (2003): Asymptotic observers of N-dimensional bilinear systems. Nonlinear Dynamics and Control, issue 3. Collection of Articles, S. V. Emelyanov and S. K. Korovin, eds. (in Russian). Fizmatlit, Moscow. 19–26.

[25] Korovin, S.K., Fomichev, V.V. (2004): Asymptotic observers for some classes of bilinear systems with a linear input. Dokl. Ros. Akad. Nauk, Control Theory, **398**, 1, 38–43.

[26] Korovin, S.K., Fomichev, V.V., Medvedev, I.S. (2005): On the minimization of the order of a fundamental observer. Dif. Uravn., **41**, 8, 1148.

[27] Korovin, S.K., Fomichev, V.V., Medvedev, I.S. (2005): Synthesis of minimal functional observers. Dokl. Ros. Akad. Nauk, Control Theory, **404**, 3, 316–320.

[28] Korovin, S.K., Fomichev, V.V., Medvedev, I.S. (2006): Minimal order functional observers. Nonlinear Dynamics and Control, issue 5. Collection of Articles, S. V. Emelyanov and S. K. Korovin, eds. (in Russian). Fizmatlit, Moscow, 51–70.

[29] Korovin, S.K., Medvedev, I.S., Fomichev, V.V. (2006): Fundamental observers for linear uncertain stationary dynamical systems. Dokl. Ros. Akad. Nauk, Control Theory, **411**, 1, 316–320.

[30] Krut'ko, P.D. (1998): Inverse Problems of Dynamics of Controllable Systems. Linear Models (in Russian). Nauka, Fizmatlit, Moscow.

[31] Roitenberg, Ya.N. (1992): Automatic Control (in Russian). Nauka, Fizmatlit, Moscow.

[32] Smagina, E.M. (1990): Problems of Analysis of Linear Multidimensional Objects with the Use of the Notion of the Zero of a System (in Russian). Tomsk Inst. Publ., Tomsk.

[33] Fomichev, V.V., Medvedev, I.S. (2004): Construction of functional observers for uncertain systems. Dif. Uravn., **40**, 8, 1146–1147.

[34] Aguilar, R., Martinez-Guerra, R., Maya-Yescas. R. (2003): State estimation of partially unknown nonlinear systems: a class of integral high gain observers. IEEE Proc. Control Theory and Applications, **150**, 3, 240–244.

[35] Aguilar-Lopez, R. (2003): Integral observers for uncertainty estimation in continuous chemical reactors: algebraic differential approach. Chemical Engineering J., **93**, 2, 113–120.

[36] Basile, G., Marro, G. (1969): On the observability of linear time-invariant systems with unknown inputs. J. of Optimization Theory Application, **3**, 410–415.

[37] Bastin, G., Gevers, M. (1998): Stable adaptive observers for nonlinear time varying systems. IEEE Trans. on Automatic Control, **33**, 7, 650–658.

[38] Bhattacharyya, S.P. (1978): Observer design for linear systems with unknown input. IEEE Trans. on Automatic Control, **23**, 483–484.

[39] Chang, J.L. (2003): Robust sliding-mode control with disturbance attenuation using only output feedback. JSME Int. J. Series C, Mechanical Systems Machine Elements and Manufacturing, **46**, 1, 239–244.

[40] Corless, M., Tu, J. (1998): State and input estimation for a class of uncertain systems. Automatica, **34**, 757–764.

[41] Darouach, M., Zasadzinski, M., Xu, S.J. (1994): Full-order observers for linear systems with unknown inputs. IEEE Trans. on Automatic Control, **39**, 3, 606–609.

[42] Dochain, D. (2003): State and parameter estimation in chemical and biochemical processes: a tutorial. J. of Process Control, **13**, 8, 801–818.

[43] Dochain, D. (2003): State observers for processes with uncertain kinetics. Int. J. of Control, **76**, 15, 1483–1492.

[44] Doyle, J.C., Stein, G. (1979): Robustness with observers. IEEE Trans. on Automatic Control, **24**, 4, 607–611.

[45] Edwards, C., Spurgeon, S.K., Hebden, R.G. (2003): On the design of sliding mode output feedback controllers. Int. J. of Control, **76**, 9–10, 893–905.

[46] Engell, S., Konik, D. (1986): Zustandsermittlung bei unbekanntem Eingangssignal. Automatisierungstechnik, **34**, 1, 38–42.

[47] Fairman, F.W., Hinamoto, T., Hirschorn, R. (1982): Disturbance decoupled observers having unrestricted spectrum. J. of Franklin Institute, **313**, 123–133.

[48] Fairman, F.W., Mahil, S.S., Luk, L. (1984): Disturbance decoupled observer design via singular value decomposition. IEEE Trans. on Automatic Control, **29**, 84–86.

[49] Ficklscherer, P., Müller, P.C. (1985): Robuste Zustandsbeobachter für lineare Mehrgrößenregelungssysteme mit unbekannten Störeingängen. Automatisierungstechnik, **33**, 173–179.

[50] Green, M., Limebeer, D.J.N. (1995): Linear Robust Control. Prentice Hall.

[51] Guan, Y.-P., Sail, M. (1991): A novel approach to the design of unknown input observers. IEEE Trans. on Automatic Control, **36**, 5, 632–635.

[52] Hermann, R., Krener, A. (1977): Nonlinear controllability and observability. IEEE Trans. on Automatic Control, **22**, 5, 728–740.

[53] Hostetter, G., Meditch, J.S. (1973): Observing systems with unmeasurable inputs. IEEE Trans. on Automatic Control, **18**, 3, 307–308.

[54] Hou, M., Müller, P.C. (1992): Design of observers for linear systems with unknown inputs. IEEE Trans. on Automatic Control, **37**, 6, 871–875.

[55] Hou, M., Müller, P.C. (1994): Disturbance decoupled observer design: a unified view point. IEEE Trans. on Automatic Control, **39**, 6, 1338–1344.

[56] Hou, M., Müller, P.C. (1994): Fault detection and isolation observers. Int. J. of Control, **60**, 5, 827–846.

[57] Hou, M., Pugh, A.C., Müller, P.C. (1999): Disturbance decoupled functional observers. IEEE Trans. on Automatic Control, **44**, 2, 382–385.

[58] Hui, R., Zak, R. (1993): Robust output feedback stabilization of uncertain dynamic systems with bounded controllers. Int. J. of Robust and Nonlinear Control, **3**, 115–132.

[59] Imai, H., Akashi, H. (1981): Disturbance localization and pole shifting by dynamic compensation. IEEE Trans. on Automatic Control, **26**, 1, 226–235.

[60] Isidori, A. (1995): Nonlinear control systems. Springer-Verlag, London.

[61] Jiang, B., Staroswiecki, M. (2002): Adaptive observer design for robust fault estimation. Int. J. of Systems Science, **33**, 9, 767–775.

[62] Kailath, T., Sayed, A.H., Hassibi, B. (2000): Linear Estimation. Prentice Hall, Upper Saddle River, NJ.

[63] Kalman, R.E. (1963): Mathematical Description of Linear Systems. SIAM J. Control, **1**, 152–192.

[64] Kalman, R.E. (1968): Lectures on Controllability and Observability. C.I.M.E., Bologna.

[65] Kimura, H. (1975): Pole assignment by gain output feedback. IEEE Trans. on Automatic Control, **8**, 509–516.

[66] Kobayashi, N., Nakamizo, T. (1982): An observer design for linear systems with unknown inputs. Int. J. of Control, **35**, 605–619.

[67] Koshkouei, A.J., Zinober, A.S.I. (2004): Sliding mode state observation for non-linear systems. Int. J. of Control, **77**, 2, 118–127.

[68] Kreisselmeier, G. (1977): Adaptive observers with exponential rate of convergence. IEEE Trans. on Automatic Control, **22**, 1, 2–8.

[69] Krener, A.J., Respondek, W. (1985): Nonlinear observers with linearizable error dynamics. SIAM J. on Control and Optimization, **23**, 2, 197–216.

[70] Kudva, P., Viswanadham, N., Ramarkrishna, A. (1980): Observers for linear systems with unknown inputs. IEEE Trans. on Automatic Control, **41**, 113–115.

[71] Kurek, J.E. (1983): The state vector reconstruction for linear systems with unknown inputs. IEEE Trans. on Automatic Control, **28**, 12, 1120–1122.

[72] Kwon, S., Chung, W.K. (2003): Combined synthesis of state estimator and pertubation observer. J. of Dynamic Systems Measurement and Control, Trans. of the ASME, **125**, 1, 19–26.

[73] Lee, C.Y., Lee, J.J. (2004): Adaptive control for uncertain nonlinear systems based on multiple neural networks. IEEE Trans. on Systems, Man, and Cybernetics, Part B, **34**, 1, 325–333.

[74] Lin, H., Zhai, G., Antsaklis, P.J. (2003): Set-valued observer design for a class of uncertain linear systems with persistent disturbance and measurement noise. Int. J. of Control, **76**, 16, 1644–1653.

[75] Lin Sheng-Fuu, Wang An-Ping (2000): Design of observers with unknown inputs using eigen-structure assignment. Int. J. of Systems Science, **31**, 6, 705–711.

[76] Luenberger, D.G. (1963): Determining the state of linear system with observers of low dynamic order. Ph.D. dissertation. Stanford University.

[77] Luenberger, D.G. (1966): Observers for multivariable systems. IEEE Trans. on Automatic Control, **11**, 190–197.

[78] Luenberger, D.G. (1967): Canonical forms for linear multivariable systems. IEEE Trans. on Automatic Control, **12**, 290–293.

[79] Mareels, I., Polderman, J. (1996): Adaptive Systems. An Introduction. Birkhäuser, Boston.

[80] Marino, R., Santosuosso, G., Tomei, P. (2001): Robust adaptive observers for nonlinear systems with bounded disturbances. IEEE Trans. on Automatic Control, **46**, 967–972.

[81] Marino, R., Santosuosso, G.L., Tomei, P. (2002): Adaptive compensation of multiple sinusoidal disturbances with unknown frequencies. Nonlinear and Adaptive Control (A. Zinober and D. Owens, eds.). Springer, 207–226.

[82] Marquez, H.J. (2003): A frequency domain approach to state estimation. J. of Franklin Institute, **340**, 2, 147–157.

[83] Miller, R.J., Mukundan, R. (1982): On designing reduced order observers for linear time-invariant systems subject to unknown inputs. Int. J. of Control, **35**, 183–188.

[84] Nijmeijer, H., Van der Schaft A.J. (1990): Nonlinear Dynamical Control Systems. Springer-Verlag, New Jork.

[85] Rosenbrock, H.H. (1970): State-Space and Multivariable Theory. Nelson, London.

[86] Rosenbrock, H.H. (1973): The zeros of a system. Int. J. of Control, **18**, 2, 297–299.

[87] O'Reilly, J. (1983): Observers for Linear Systems. Academic Press, London.

[88] Park, J.-H., Park, G.-T. (2003): Adaptive fuzzy observer with minimal dynamic order for uncertain nonlinear systems. IEEE Proc. Control Theory and Applications, **150**, 2, 189–197.

[89] Park, Y., Stein, J.L. (1988): Closed-loop, state and inputs observers for systems with unknown inputs. Int. J. of Control, **48**, 3, 1121–1136.

[90] Peter, K., Scholing, I., Orlik, B. (2003): Robust output-feedback h-infinity control with a non-linear observer for a two-mass system. IEEE Trans. on Industry Applications, **39**, 3, 637–644.

[91] Rajamani, R., Cho, Y.M. (1998): Existence and design of observers for nonlinear systems: Relation to distance of unobservability. Int. J. of Control, **69**, 5, 717–731.

[92] Rapaport, A., Gouze, J.L. (2003): Parallelotopic and practical observers for non-linear uncertain systems. Int. J. of Control, **76**, 3, 237–251.

[93] Roman, J.R., Bullock, T.E. (1975): Design of minimal-order stable observers for linear functions of the state via realization theory. IEEE Trans. on Automatic Control, **20**, 613–622.

[94] Saberi, A., Sannuti, P. (1988): Squaring down by static and dynamic compensators. IEEE Trans. on Automatic Control, **33**, 358–365.

[95] Saberi, A., Sannuti, P. (1989): Time-scale structure assignment in linear multivariable systems using high-gain feedback. Int. J. of Control, **49**, 6, 2191–2213.

[96] Saberi, A., Chen, B.M., Sannuti, P. (1996): H_∞-Optimal Control. Prentice Hall.

[97] Sannuti, P., Saberi, A. (1987): A special coordinate basis of multivariable linear systems—finite and infinite zero structure, squaring down and decoupling. Int. J. of Control, **45**, 1655–1704.

[98] Sayed, A.H. (2001): A framework for state-space estimation with uncertain models. IEEE Trans. on Automatic Control, **46**, 7, 998–1013.

[99] Sebald, A., Haddad, A. Robust state estimation in uncertain systems: Combined detection-estimation with incremental mesa criterion. IEEE Trans. on Automatic Control, **22**, 5, 821–825.

[100] Slotine, J., Hendricks, J., Misawa, E. (1987): On sliding observers for nonlinear systems. Trans. of the AMS. J. of Dynamic Systems, Measurement and Control, **9**, 245–252.

[101] Sundareswaran, K.K., McLane, P.J., Bayoumi, M.M. (1977): Observers for systems with arbitrary plant disturbances. IEEE Trans. on Automatic Control, **22**, 5, 870–871.

[102] Takahashi, R.H.C. (1999): Discrete-time singular observers: H_∞^2 optimality and unknown inputs. Int. J. of Control, **72**, 6, 481–492.

[103] Tan, C.P., Edwards, C. (2003): Sliding mode observers for robust detection and reconstruction of actuator and sensor faults. Int. J. of Robust and Nonlinear Control, **13**, 5, 443–463.

[104] Trinh, H., Ha, Q.P. (2000): Design of linear functional observers for linear systems with unknown inputs. Int. J. of Systems Science, **31**, 6, 741–749.

[105] Tsui, C.C. (1996): A new design approach to unknown input observers. IEEE Trans. on Automatic Control, **41**, 3, 464–467.

[106] Tuel, W.G. Jr. (1967): An improved algorithm for the solution of discrete regulation problem. IEEE Trans. on Automatic Control, **12**, 522–528.

[107] Utkin, V.I., Guldner, J., Shi, J. (1999): Sliding Mode Control in Electromechanical Systems. Systems and Control Series. Taylor and Francis.

[108] Valcher, M.E. (1999): State observers for discrete-time linear systems with unknown inputs. IEEE Trans. on Automatic Control, **44**, 2, 397–400.

[109] Walcott, B.L., Zak, S.H. (1987): Comparative study of nonlinear state-observation techniques. Int. J. of Control, **45**, 2109–2132.

[110] Wang, Z., Unbehauen, H. (2000): A class of nonlinear observers for discrete-time systems with parametric uncertainty. Int. J. of Systems Science, **31**, 1, 19–26.

[111] Wonham, W.M. (1974): Linear Multivariable Control: a Geometric Approach. Lecture Notes in Economics and Mathematical Systems. Springer, Berlin.

[112] Xiong, Y., Saif, M. (1998): Sliding Mode Functional Observers and its Application for Robust Fault Diagnosis. Technical Report. Simon Eraser University.

[113] Xiong, Y., Saif, M. (2003): Unknown disturbance inputs estimation based on a state functional observer design. Automatica, **39**, 8, 1389–1398.

[114] Yan Xing-Gang, Lam, J., Xie, L. (2003): Robust observer design for non-linear interconnected systems using structural characteristics. Int. J. of Control, **76**, 7, 741–746.

[115] Yang, F., Wilde, R.W. (1988): Observers for linear systems with unknown inputs. IEEE Trans. on Automatic Control, **33**, 7, 677–681.

[116] Yiiksel, Y.O., Bongiorno, J.J. (1971): Observers for multivariable systems. IEEE Trans. on Automatic Control, **16**, 603–621.

[117] Youssouf, A., Kinnaert, M. (1977): Partial state estimation in presence of unknown inputs. European Control Conference (Brussels), 1–4 July.

[118] Zak, S.H., Hui, S. (1993): Output feedback in variable structure controllers and state estimation of uncertain nonlinear dynamical systems. IEE Proc. Control Theory and Applications, **140**, 41–50.

[119] Zhou, K., Doyle, J. (1998): Essentials of Robust Control. Prentice Hall, New York.

[120] Zhou, K., Doyle, J.C., Glover, K. (1996): Robust and Optimal Control. Prentice Hall.

[121] Zinober, A., Owens, D. (2001): Nonlinear Adaptive Control. Lecture Notes in Control and Information Sciences. Springer-Verlag.

[122] Darouach, M. (2000): Existence and design of functional observers for linear systems. IEEE Trans. on Automatic Control, **45**, 5, 940–943.

[123] Trinh, H., Tran, T.D., Nahavandi, S. (2006): Design of scalar functional observers of order less than $(\nu - 1)$. Int. J. of Control, **79**, 12, 1654–1659.

Index

Bei Fragen zur Produktsicherheit wenden Sie sich bitte an:
If you have any questions regarding product safety,
please contact:

Walter de Gruyter GmbH
Genthiner Straße 13
10785 Berlin
productsafety@degruyterbrill.com